The Quantum Light War

Abigail Wong, Panav Kalra, Tony Lim

Table of Contents:

Chapter 1: Introduction   3

Chapter 2: The History of Photonics   19

Chapter 3: Technical Information   49

Chapter 4: Biophonics (medicine)   63

Chapter 5: Other Prominent Fields   121

Final Words   144

References   145

Chapter 1: Introduction

Light is everywhere. Without it, we would not be able to see. Therefore, it is no surprise that humans have studied light since as early as the ancient Greeks. From the very beginning, the ability to manipulate light has helped us progress further. Starting from the simple initial discovery of fire, humans have continued to uncover many new, pioneering secrets about the power of light for centuries—from more practical uses such as the invention of the telescope and the creation of the lightbulb to more complex theoretical revelations like the discovery of light waves, and finally, the discovery of photons.

Our advancements in the field of light have given birth to the concept of a light quantum (now known as a photon). This idea was first hypothesised by Max Planck and advanced by Albert Einstein, who proposed the idea of a light quantum—a quantum being the individual building block of anything that exists, including light—to explain the photoelectric effect discovered by Heinrich Hertz in 1887. This is the emission of electrons when electromagnetic radiation, such as light, hits a material. During his experiment, Hertz observed that sparks shoot when ultraviolet light shines onto a metal plate [1]. Since electrons hadn't been discovered yet, the relationship between electricity and light was only clarified later by Philipp Lenard, a different German physicist, in 1902. He demonstrated how light causes electrons to emerge from a metal surface[1]. However, the puzzling part was not the emission itself, but the fact that increasing the light intensity produced more electrons instead of increasing their energy. On the other hand, increasing the frequency of the light produced electrons with higher energies with no increase in the number produced.

The problem is that if light is merely an energy-carrying wave, then it is impossible to comprehend the photoelectric effect. One reason is that if light were considered a wave, increasing the intensity or energy of the wave should lead to electrons absorbing more energy and emerging with more energy, due to the principle of energy conservation stating that energy cannot be created or destroyed, and that the amount of momentum in a system should remain constant from the principle of momentum conservation. However, another explanation was needed because that's not what occurred in the experiment. Einstein explained that a light quantum transfers all its energy to a single electron in the material and disappears[2]. This extra energy from the light quantum, or photon, allows the electron to emerge from the metal surface and leave. By considering light as particles and that increasing its intensity meant increasing the number of photons of the same energy—rather than increasing the total energy of a wave—the phenomenon of growing electron emission but not energy can be explained.

This concept begs the question of whether light is a wave or a particle. Some scientists debated over this, but the theory of light being a particle eventually vanished. That is, until Albert Einstein revived it at the end of the 19th century. Now, thanks to the double-slit experiment, among the most well-known physics experiments, there is compelling evidence that energy and matter, including light, can manifest characteristics of both waves and particles—termed as the particle-wave duality of matter[3]. Imagine a light shining at a wall with two slits in it. Some of the light will pass through the slits, and this causes the light to split into two waves. When a wave's peak—its highest point—meets a trough at certain points, the lowest section, they will cancel out one another. Conversely, they reinforce one another at other spots where peak hits peak, creating a brighter light. Imagine that behind the first wall there is a second wall: you might think that the light passing through the slits of the first wall will form two rectangular segments on the second wall, as you would imagine particles like sand would behave if you shot them through the two slits. However, the results of this experiment show that when the light meets the second wall located behind the first, multiple bright lines will form due to parts of the waves intensifying each other while others cancel out, a phenomenon referred to as an interference pattern[4]. When two waves meet and interact, their intensities or amplitudes get added together. Constructive interference is when two waves meet in phase (e.g., peak meets peak or trough meets trough) so their intensities or amplitudes are combined, creating the bright strips. Destructive interference, in contrast, occurs when two waves are completely out of phase (e.g., peak meets trough) so when the intensities are added, they cancel each other, creating the dark strips.

On the other hand, in theory, since electrons, for example, are particles, this phenomenon shouldn't happen if they were fired through the slits. It has been tested that if you block one of the slits and fire electrons through it, the pattern the electrons form on the second wall is indeed just one rectangular strip[4]. However, strangely, if you open the second slit and shoot electrons through both, the areas where electrons collide reproduce a wave's interference pattern[4]. Hence, this experiment shows that "particles" such as electrons can exhibit behaviours of both waves and particles, suggesting that light and photons are the same. In simplified terms, light can be understood to be made up of particles known as photons, while a light wave is the flow of these photons. This can be analogous (comparable) to a wave in the ocean, with a wave being composed of smaller particles like water droplets or sand.

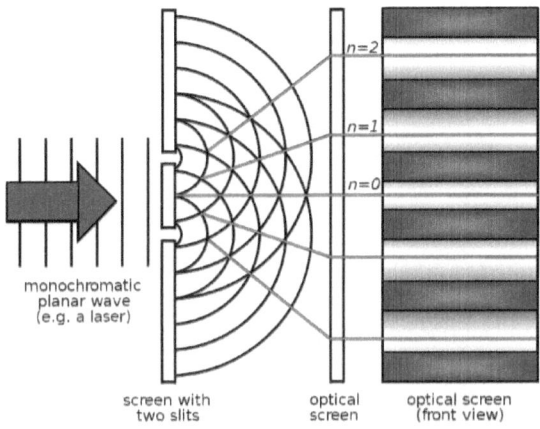

Diagram showing the setup of the double-slit experiment using light waves[1]

So, what is a photon exactly? The word "photon" is derived from the Greek word "phos," meaning light. As follows, photons are massless particles that carry electromagnetic force[5]; in other words, they are the quantum of light or other electromagnetic radiation. A photon carries energy proportional to the radiation frequency—the number of waves that pass a given point in one second. This means that the more times a wave passes a given point, the more energy the photon carries. Photons are constantly in motion and have a constant speed of light in a vacuum[6].

Incidentally, although the words "photonics" and "optics" are often used synonymously, there is a distinction. Optics, the study of light and its use in technical instruments, has been around for a long time, centering on the eye and vision at first but then advancing to a broader range of possibilities with the inventions of more optical instruments. Conversely, the field of photonics only emerged relatively recently. It was initiated in the 1960s with the invention of lasers, which remain the bedrock of many photonics applications. Photonics highlights the dual nature of light in quantum mechanics and is better described as a subset of the optics discipline which focuses explicitly on the use of photon-based light creation, detection, and manipulation, mainly for tasks that were formerly considered to be within the conventional scope of electronics.

Even before lasers, there was a lesser-known precursor known as the maser. In 1954, Charles Hard Townes, Herbert J. Zeiger, and James P. Gordon demonstrated the first maser at Columbia University. A maser is similar to a laser except for the fact that it produces and amplifies radiation in the microwave frequency, which has a longer wavelength or lower frequency, meaning humans can't see it. Longer wavelength refers to a greater distance from one maximum or minimum point of a wave to another, and lower frequency means fewer waves pass through one point in a second. Atomic clocks use masers as their timekeeping mechanism. This creation would later inspire and progress towards the invention of lasers.

But Townes saw that the shorter wavelengths of optical and infrared light would be capable of creating even more powerful tools. When visiting Schawlow, another physicist at Bell Labs, he mentioned expanding the master principle to Schawlow[7]. Schawlow suggested setting up a series of mirrors, which turned out to be a crucial part of lasers and whose function will be explained later. A proof-of-concept paper on their work was written and published by the two men in the Physical Review edition from December 1958, several months after their first work. Interestingly, it was two years later when not only did they obtain a patent for the laser's invention, but the first functional laser was constructed that same year.

On May 16, 1960, Theodore Maiman operated the first laser at the Hughes Research Laboratory in California by using a powerful flash lamp to shine on a ruby rod whose surfaces were covered in silver. Immediately, he submitted a report of the work to the scientific journal Physical Review Letters. However, the editors of the journal declined to publish the report. There has been speculation that the reason for this was the announcement that the journal had made. It had declared that any more documents concerning masers would be rejected due to the sheer number of such papers being submitted[8]. However, Simon Pasternack, an editor of Physical Review Letters during that period, has stated that it was because of Maiman's earlier publication in the same year of another study on the excitation of ruby with light, a concept explained later, that he declined this historic paper[8]. In his opinion, the new work appeared to be just a repetition of the same concept[8]. Pasternack's response illustrates how little was known about lasers and their importance at the time[9].

Determined to publicise his results, Maiman turned to Nature for help, and when public announcements about the publication process were underway, quite a stir was created. Because the full Nature research had not yet been formally published, there was scepticism among scientists over the front-page newspaper speculations of potential "death rays."[9]. In this initial laser, Maiman excited, or supplied energy to, the ruby's electrons using a pulsing light source with merely a few milliseconds between each pulse[9]. Thus, the laser only emitted a brief burst of light instead of a continuous wave. Despite the short duration, this significant energy release surpassed initial expectations and exceeded the power levels envisioned in earlier discussions about lasers. Before long, in the same laboratory, "Q switching" was implemented, a technique where the laser light pulse was further shortened and its instantaneous power in watts was increased to extend beyond the millions[9]. Nowadays, the output of a laser can reach a million billion watts. To put that into perspective, in an aircraft like a Boeing 777 with two engines, each engine produces roughly 23 megawatts or 23 million watts of power.

Just the following year, lasers became commercially available, the optical fibre laser was invented, and there was the first medical use of a laser on a human patient—a ruby laser from American Optical was used to destroy a retinal tumour. In 1969, the use of lasers made headlines. The Apollo 11 crew left a laser ranging retroreflector (LRRR), similar to a mirror, on the surface of the moon. Apollo 11 was the spaceflight that carried the first humans to the moon, including Neil Armstrong, the first man to step on the moon. During this mission, laser beams were shot from Earth to the moon and reflected back by the LRRR, which was used to give an exact measurement for the average distance from the Earth to the moon as well as other data.

Lasers may sound like something out of science fiction, but in reality, lasers can be found everywhere in daily life. An uncommonly known fact about the word "laser" is that it is, in fact, an abbreviation for Light Amplification by Stimulated Emission of Radiation[10]. As the name implies, lasers work when energy from an electrical current or light stimulates electrons in the atoms of optical materials like glass, crystal, or gas, causing the electrons to absorb and amplify the energy until a laser beam is generated. The electrons are sufficiently "excited" by the additional energy, enough to move their orbit around the atom's nucleus from one of lower energy to one of higher energy. This happens because an atom contains a nucleus and a "cloud" of electrons spiralling around the nucleus. Atoms can exist in different excitation states, meaning they can possess different energies. In simplified terms, if we provide energy to an atom, which could come in the form of heat, we might anticipate that some of the electrons in the lower-energy orbitals would move into higher-energy orbitals which are further from the nucleus. However, an electron ultimately seeks to return to its ground state after entering a higher-energy orbit—the lowest possible energy level of the atom when it is not charged or excited by external sources[11]. When this occurs, energy is released as a photon[11].

To better picture this effect, consider when metal turns red when heated for a long time. The red colour is caused by electrons, stimulated by heat and releasing radiation or photons with wavelengths including those which produce red light. Any source that generates light accomplishes this by the process of electrons transitioning between different energy levels, resulting in the emission of photons. A laser is simply a device that modulates the release of photons from energised atoms[11]. In general, highly intense bursts of light or electrical beams are used to stimulate the laser medium, creating a sizable cluster of atoms in the excited state within the medium[11]. Another critical component of lasers is mirrors. After the electrons release photons, these photons go back and forth across the laser medium by reflecting off the mirrors. Some of the reflected photons strike more atoms, increasingly generating photons which continue to bounce between the rod's internal mirrors[12]. There is a "half-silvered" mirror located at one end of the laser, meaning it only reflects some light while letting others through[11]. When the photons are sufficiently amplified to pass through the optical substance and mirror simultaneously, the laser light is able to emerge [12].

Although lasers can be thought to be sources of light, how does laser light differ from a regular flashlight? Firstly, only monochromatic light is released in a laser; this means only one specific colour or wavelength of colour is released[11]. The energy that is discharged as an electron descends to a lower orbit determines the wavelength[11]. Secondly, coherency is another property exhibited by the released light, meaning that every photon has equal frequency and launches its wave fronts synchronously[11]. Imagine if there were waves in the sea all parallel to each other and having peaks at the same position, or in technical terms, in the same phase. Consequently, each wave amplifies the others, creating a larger, or in this case, brighter wave. Also, laser light is extremely directional as a laser beam is incredibly focused and powerful, having a very narrow beam[11]. In contrast, a light source such as a torch emits light over a wider area in various directions, at a very low intensity with a diffuse quality.

These properties can be explained through the "stimulated emission" part of the acronym "laser," another phenomenon predicted by Albert Einstein in 1916. In contrast to a flashlight, where photon emission from atoms occurs randomly, stimulated emission ensures a more organised release of photons. As mentioned, a particular wavelength is associated when a photon of any atom is released[11]. Stimulated emission works similarly to the process explained earlier where an incoming photon may engage with an excited electron, inducing its transition to a level of lower energy as a result. However, the difference is it can only occur when the incoming photon with the specific wavelength encounters a different atom with an electron that shares the incoming photon's excited state, such that the photon later emitted by the atom oscillates with identical frequency and direction[11].

It should be noted that lasers are not limited to one type; various lasers exist. Examples include solid-state lasers, gas lasers, diode or semiconductor lasers, dye lasers, and others. The distinction between the types is primarily the laser medium, the material used to produce the stimulated emission. For example, gas lasers use gases such as carbon dioxide and helium as the laser medium. Each of them serves a distinct purpose and can produce different wavelengths. Most lasers have overlapping applications like in medicine for laser surgery or spectroscopy. However, there are applications specific to each laser, like how solid-state lasers are used in LIDAR technology. On the other hand, dye lasers have the ability to produce a much broader wavelength range, making them capable of being tunable lasers, meaning that their wavelength can be adjusted in real time while in operation. An instance of this being a beneficial property is in laser isotope separation where the lasers can be tuned to specific atomic resonances in order to separate the isotopes. The differences in properties can be seen in laser marking, for example. This is the process of marking parts or components using laser technology to leave permanent marks on the material. The wavelengths produced by solid-state lasers are better suited for marking metal, while organic materials react better with the wavelengths produced by gas-state lasers. This demonstrates the versatility of lasers and how different types can be used for different advantages to cover a wide range of applications.

Indeed, lasers have played a fundamental role in many photonics applications, including in optical disk drives and glass-fibre-optic communications that transmit laser signals carrying data at speeds potentially up to 40 gigabits (billions of bits) per second. This technology has been so successful that by the end of 2018, the laser market was estimated to be worth more than 12.9 billion USD. At Stanford University, a more efficient laser beneficial for the trend of the time was developed. Smaller chips with lower manufacturing costs were favoured as they were also more manageable to put into smaller electronic devices since it was possible for about

400,000 of those lasers to fit into a one-centimetre-square chip. According to the researchers who built the laser, their invention's theoretical capabilities surpass the speed of 100 gigabits per second, which is 2.5 times faster than the communications industry's target of 40 gigabits per second. There are still limiting factors though, such as transmission delays and heat build-up on the chip. Nonetheless, this is just one example of how photonics has advanced technology.

Early photonics was predominantly focused on telecommunications applications. In 1966, Charles Kao, a Chinese physicist known as the father of fibre optics communications, revealed how to make improved fibre suitable for communications by using an optical cladding, a layer wrapped around the glass core material, as well as by removing impurities. Following this revelation, the term "photonics" gained widespread usage in the 1980s with the adoption of fibre-optic data transmission by telecommunications network providers.

Fibre optics is a particularly important area of photonics and is a commonly mentioned topic. How it actually works may be less understood though. Fibre optic cables work with the concept of total internal reflection. This is the total reflection of a light beam within a material, like glass or water, meaning the ray enters the medium and reflects on the edge back within the medium. The phenomenon happens when light strikes a surface at an incidence angle (angle between the light ray and an imaginary line perpendicular to the surface) greater than a specific limiting angle known as the critical angle. This occurs when the refractive angle, the angle at which the ray bends after passing a surface, exceeds 90 degrees so it reflects back within the medium.

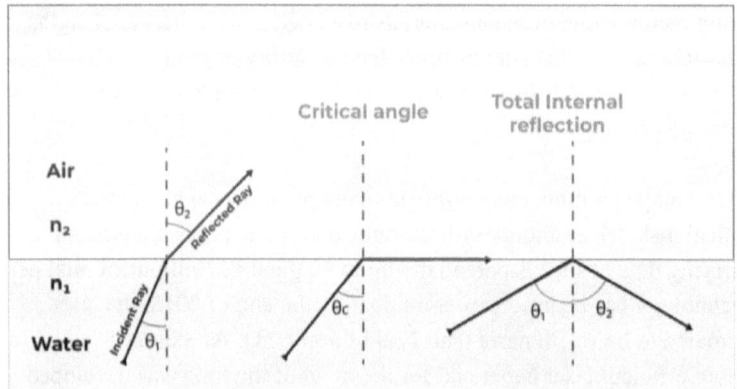

Illustration of the angles during total internal reflection versus normal refraction[2].

Optical cladding is important because refraction, the bending of light, only occurs when light passes through media of different densities, such as from water to air, due to its shift in speed of travel. There is a property known as the refractive index, which establishes how quickly light moves through a medium[13]. Different materials have different refractive indices. When light passes from a medium with a higher refractive index to one with a lower refractive index, light travels faster and bends away from the normal—an imaginary line perpendicular to the surface the light passes through. In other words, the light bends more in the direction back towards the medium. Hence, using a glass core with a higher refractive index than the optical cladding causes total internal reflection, allowing light to be confined within the cable.

Optical fibres are similar to computers and other digital devices in the sense that they process and store information in terms of ones and zeros. Every file you access is composed of a large number of ones and zeros, called bits. When data is transferred between places, those bits are transmitted one by one along either copper wiring, the air, or optical fibre for situations requiring long-distance communications[14]. For example, with phone lines—which are actually copper cables—the information is carried as electric charges travelling through the cable[14]. The presence of an electric charge signifies the transmission of a value of 1, while the absence of an electric charge signifies the transmission of a value of 0. By quickly switching the current on and off, at least millions of times every second, the bits contained in the file travel along the cable[14]. Similarly, in optical fibre cables, on one end, a laser will pulse on and off. This travels to the opposite end of the cable where there is a light sensor that records a 1 in the presence of light and a 0 in the absence of light[14].

Fibre optics is one of the most famous examples of where photonics is used. However, its usage extends to a plethora of other industries and fields, revolutionised by photonic advancements. Examples include artificial intelligence, biological and chemical sensing, medical diagnostics and therapy, data centres, and optical computing[15]. In fact, photonics can even be seen in daily life in much of the technology that utilises light, such as smartphones, televisions, laser printers, and LED lighting.

Telecommunications is the main application of fibre optics. This refers to the technology that internet services employ to send light pulses carrying information across great distances through glass or plastic fibre strands. Download and upload speeds have increased dramatically since the implementation of optical fibre. Cloud computing centres from Facebook, Amazon, and Google, for example, are receiving continuously increasing amounts of data traffic. The use of integrated photonics systems will allow them to cope with the terabit-scale data rates of traffic with nanosecond switching speeds while saving costs by only needing to consume half as much power[16].

Although the advancements in photonics have improved our quality of life tenfold, how did people know to invest so much in photonics even in its early stages? Back in 1948, through the invention of the first transistor—and the first commercially produced microprocessor, the Intel 4004 of 1971—computer microprocessors have gotten progressively faster and more powerful. This rapid advancement can be attributed to the fact that transistors, the basic building blocks of modern technology, have been doubling in number on integrated circuits about every two years[17], as observed by Gordon Moore in 1965 and known as Moore's Law.

Transistors are among the fundamental base components of much modern technology. They are semiconductor devices used to switch or magnify electrical impulses and power. The electrical conductivity of semiconductors is in the range between that of an insulator like glass and a conductor like metals. They are used extensively for computing or electrical-related industries because of this property. With increasing temperature, while metals' electrical conductivity exhibits a reduction in conductivity, a semiconductor's electrical conductivity actually increases.

Impressively, in 2022, IBM (the International Business Machines Corporation, a multinational technology corporation based in the United States known as "Big Blue") announced that it had successfully developed a two-nanometer technology which is barely ten times larger than an atom of silicon. This would make it possible to fit 50 billion transistors into a fingernail-sized chip[18]. However, as the transistors get smaller and smaller, they will eventually reach their physical limit, where another method to improve the transistors is needed. This is where photonics could come into play.

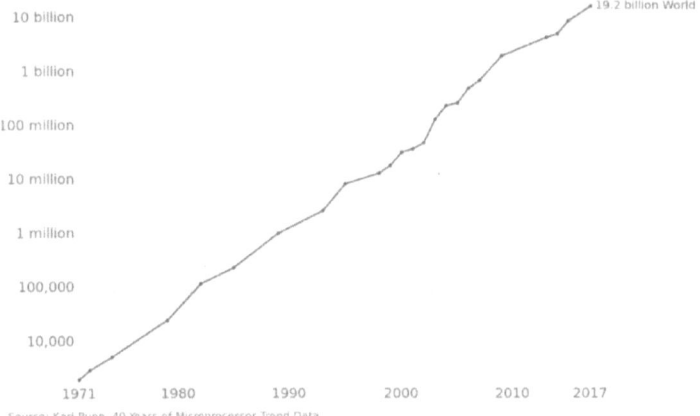

A semi-log plot of transistor counts for microprocessors against dates of introduction, illustrating that it is nearly doubling every two years[3]

How does photonics differ from the electronics we've all known and used? It is without a doubt that photonic devices have advantages over other existing technologies such as traditional electrical systems. In order for electrical components to be active, they need an electrical supply to be connected via a switch. On the other hand, a light source, most likely a laser, is used to activate photonics components[19]. Since photonics utilises the movement of photons from light waves through a circuit rather than electrons, the data transmission can happen close to the speed of light, just like in fibre optic cables. It is only close due to the fact that since the light is not travelling in a vacuum, the full speed of light cannot be reached. Even so, being able to travel nearly at the speed of light, photons can still move more rapidly and more efficiently since they are virtually frictionless[19]. In contrast, for electronic components, electrons experience resistance due to them engaging with other particles, slowing them down when travelling through copper wires and generating heat which then leads to energy loss[19]. This means that compared to all-electrical circuits, this technology can deliver high-speed, high-bandwidth, energy-efficient, and integrated solutions that can and are already transforming the way our current technology works.

The promising future of photonics has led to hundreds of billions of dollars being poured into the field. According to Future Market Insights, in 2023, around 911.5

million USD was invested, and that number is only expected to grow well past a trillion USD by 2030. Many companies, big or small, have invested in this technology. For example, PsiQuantum is the name of a company that designs and manufactures large-scale, distributed, fault-tolerant quantum computers. In 2023, PsiQuantum announced it raised 100 million USD in a funding round led by Microsoft and other investors. They claimed that Microsoft and PsiQuantum will work together to power the global quantum ecosystem. Many start-ups and companies have emerged as well with the growing importance of the photonics field. In 2021, Rockley Photonics, "a global leader in photonics-based health monitoring," went public. They are committed to exploring novel approaches for measuring critical biometrics with various photonic integrated circuits (PIC), sensors, and full-stack solutions[20].

Although there remain many difficulties in creating an entirely light-based computer chip where all logic is done using optical components, there exists an example of a more practical use of photonics currently through the use of optoelectronics. The function of these devices is to encode digital information into electromagnetic signals and convert electromagnetic signals back into electrical ones. These components are typically physically separate from one another and need to be integrated together by coupling—meaning transferring electrical energy from one part of a circuit to another—to create a complete optical circuit.

Integrated photonics (IP) emerges as a developing field within photonics. In IP, photonic devices and waveguides, a component elaborated on later, are manufactured directly on a flat substrate composed of materials like Indium Phosphide (InP) or Silicon Nitride (SiN). This innovation means that, similar to electrical signals being conducted in electronic integrated circuits, light can now be processed and transmitted by photonic integrated circuits[19]. These technologies can be seen as analogous to each other at times. For example, waveguides in photonic integrated circuits are used to control and direct light, comparable to the way that wires can transmit electrical signals[19].

As mentioned above, silicon is a common material used in computing for computer chips, transistors, and other electronic devices. Over the years, researchers have continued to develop more ways to incorporate silicon into different devices to improve their functionalities. The first monolithic integrated circuit was invented by Robert Noyce in 1959, made from silicon. It also had a crucial role in photonics, with one of the leading areas of research within technology in general being silicon photonics. Many believe that the growing advancements in silicon photonics will boost the photonics field tremendously.

Silicon photonics merges electronics and photonics into one field of study as silicon is able to integrate all components on a single integrated chip. This includes optical components, electronic circuits, and even microelectromechanical systems, making it extremely suitable for making photonic integrated chips. It can be thought of as a similar concept to fibre optics cables except it uses silicon components all in one compact PIC rather than kilometres of cables under the ocean.

Besides its abundance from being the second most common element found on Earth, silicon possesses impressive characteristics that make it ideal for a range of uses. Firstly, its basic cubic crystal structure makes it possible to produce defect-free wafers which are thin slices of semiconductor substance used in electronics for the making of integrated circuits with exceptional purity[21]. Other advantageous properties in terms of the production of semiconductor devices include its thermal conductivity, low density, and hardness[21]. Also, at specific wavelengths, silicon possesses a high refractive index meaning it can refract, or bend light more effectively. This quality is beneficial for shrinking devices to incredibly miniature scales as it enables light confinement in the submicrometer range[21]. This allows for extra precise control of light at the desired location and tight bending of optical waveguides. Furthermore, standard methods used in semiconductor or CMOS processing can be readily implemented in silicon photonics, facilitating cost-effective mass manufacture[21]. CMOS refers to both the manufacturing techniques and a specific design approach for digital circuits. It is extensively utilised in the majority of modern integrated circuits, like chips or microchips.

In addition, silicon has a high-quality native oxide, a natural compound made of two of the earth's most abundant materials: silicon and oxygen. This native oxide provides superior qualities compared to other semiconductors[21]. For example, the oxide is a great material for silicon waveguide cladding[21]. Waveguides are another common component found in photonics. Put simply, waveguides are structures that guide waves by limiting energy transfer in a single direction. Again, they can be thought to be comparable to how wires carry electricity in a circuit. One kind of waveguide is fibre optic cables. This silicon waveguide cladding is the same concept as optical cladding. Modulating the oxide cladding layer allows for the manipulation of light propagation, the process by which an electromagnetic wave transfers energy from one point to another or simply the movement of waves, within silicon waveguides[21]. This ability to manipulate light is extremely crucial in developing optical or photonic devices which utilise light for data transmission. Collectively, these properties mean silicon has provided a promising direction for photonics development.

Next, here is a brief synoptic of how photonics has helped some more industries around the world. The development of photonics through silicons and decades worth of advancements has allowed technologies, notably sensors, to be significantly impacted by the application of photonics. Using photons, we can identify and distinguish a material's optical characteristics. This means they are able to detect various things, from chemicals in gases, to contamination in water or food, to abnormalities in our body such as blood pressure or pulse. The amount of light that can travel through something or how much of it is reflected back can provide different data on what substances are in it. These sensors are even used in mobile devices for monitoring health by tracking physical activity, heart rate and sleep. These sensors have allowed for higher sensitivity and a reduction in invasive procedures in medicine for example, a pulse oximeter can replace having to retrieve a blood sample to find the oxygen levels in your blood. As a matter of fact, photonics aided in the COVID-19 pandemic disease detection to provide faster and more accurate results.

The application of photonics has also extended into a relatively new technology: artificial intelligence. Photonic integrated circuits have facilitated the creation of high-speed artificial neural networks. Neural networks are under the branch of artificial intelligence. More specifically, they are categorised under deep learning, a type of machine learning. This process starts by training computers to process information to mimic how the human brain works. This works by using artificial neurons that are connected loosely in the same manner as biological neurons that connect to the brain. With machine learning methods becoming increasingly complex and as the demand for artificial intelligence and machine learning grows in virtually every industry, algorithms running on photonic systems may be able to address the upcoming challenges. For example, the creation of neuromorphic electronics has highlighted some of the challenges in this field, especially those related to processor latency, or in other words, delay in the processor. Meanwhile, neuromorphic photonics can reduce latencies to sub-nanoseconds and increase storage capabilities, allowing for opportunities to extend A.I. abilities by improving A.I. systems' adaptability and learning capabilities. Neuromorphic photonics greatly benefits neuromorphic networks, which are better at finding patterns and predictions, along with reinforcement learning, where the machine learns from trial and error, enhancing the accuracy of machine decision-making.

In addition to laser research, LiDAR technology is another prominent example of photonics revolutionising various applications. LiDAR stands for light detection and ranging and is the technology used in self-driving cars to diagnose road conditions and avoid collisions precisely. It uses lasers and other photonics devices to detect objects in its path. Previous versions of LiDAR primarily relied on moving parts, creating many disadvantages, such as resulting in them being big, slow, expensive, low-resolution, and prone to premature failure. However, integrated photonics has fixed these problems by shrinking LiDAR technology to less than 0.1mm, scanning without moving parts, and being manufactured in large quantities cheaply[16].

Other active areas of research in photonics include quantum photonics and nanophotonics. Quantum photonics deals with the scientific study of generating, modifying, and identifying light in domains where the individual quanta, the photons within the light field, are controllable coherently[22]. Quantum photonics are applied to quantum information processing, including quantum computing, a term many people have probably heard of but may need help understanding the meaning. In comparison, nanophotonics focuses on investigating the behaviour of light on tiny scales of nanometers and the effects of light on these nano-sized objects.

Before we dive into deeper areas of quantum research and the contributions of photonics to it, let us review the fundamentals of quantum mechanics. It is a fundamental theory in physics that deals with the behaviour of matter at and below the scale of atoms, including particles such as electrons and photons. Theories such as the wave-particle duality are also covered in this field. Consequently, within quantum mechanics, the multidisciplinary discipline of quantum computing tackles complicated problems quicker than traditional computers. Instead of using bits, the fundamental unit of information that a classical computer can process and store, quantum computing uses quantum bits, which are shortened to Qubits. What makes qubits attractive compared to regular bits is that bits always exist in one of two physical states, similar to an on/off light switch, encoded as a single value, typically a 0 or 1. In contrast, Qubits can exist in a superposition, the capacity to simultaneously exist in multiple states until simultaneously measured at 0 and 1[23]. This means n Qubits can store and process as much data as 2n bits.

Because of their light-speed transmission and simplicity in manipulation, photons are particularly appealing quantum information carriers[22]. Historically, quantum photonics experiments have involved "bulk optics" technology with individual optical parts assembled onto a huge optical table, adding up to hundreds of kilos of mass[22]. Once again, one crucial step toward creating practical quantum technology is implementing photonic integrated circuit technology since photonic chips are advantageous over bulk optics in terms of several capabilities, such as miniaturisation, stability, increased experiment size, and manufacturability[22].

Photonics extends to the development of nanotechnology; this is the investigation of light behaviour at the nanometer scale, equal to one thousand-millionth of a metre, and how light interacts with nanostructures[24]. The applications range from electrical engineering to biomedicine to even carbon-free energy. Some achievements include record-setting photovoltaic efficiencies or the efficiency rates of solar panels. These ultra-sensitive sensors can identify individual molecules and minute quantities of viruses and bacteria with treatments that can kill tumours non-invasively. Nanotechnology has been a hot research topic since its discovery, and combined with photonics, it can transform almost any industry imaginable.

After a quick discussion of different fields that photonics has impacted so far and how it pushes the progress of our technology, it is evident that photonics will only continue to have a bigger role in our lives as it keeps evolving. Therefore, the rest of this book will continue to delve into further details about the fields with the most revolutionary advancements.

Chapter 2: The History of Photonics

A brief history of photonics was covered in the introduction, but more exciting stories and essential names in the photonics timeline should be discussed. As you know, photonics began when the laser was invented in the 1960s. If you think of pioneers in optics, you might think of names like Albert Einstein and Galileo Galilei. However, there are actually many other famous names from history, such as Euclid, Euler, and Leonardo da Vinci, whose contributions to optics are less widely known than their other works. Although this section will not focus on the specific names mentioned, some more contemporary contributors to photonics will be explored.

Holograms

One example of how the development of lasers advanced other areas of photonics to benefit society is the invention of holograms. Near the end of the 1940s, Dennis Gabor tried to increase the electron microscope's resolution. To do so, he devised wavefront reconstruction, now commonly known as holography[25]. To start with some background information, while Gabor attended the Berlin Technical University from 1920 to 1927, he had the chance to work with brilliant minds such as Albert Einstein and Max Planck before receiving his doctorate in engineering[25]. He then moved to England, where he carried out most of his research on the electron microscope. Here, he focused on enhancing the resolving power, the capacity to tell two lines or points apart in an object[25]. He came up with a potential solution in 1947. Gabor planned to use optical techniques to clarify a distorted electron image. His intended approach involved splitting an electron beam in two. Among the subsequent beams, one would aim at the target image object while the other aimed towards a mirror[25]. Despite their initial coherence characterised by their identical frequencies and wavelength oscillation of photons, interference patterns would appear on a photographic film as the two beams converged because of how distinct surfaces reflect them[25].

Consequently, when light was projected through the film, it generated a three-dimensional reconstructible image. In simpler words, instead of taking a direct picture of the subject as usual when looking through microscopes, the photosensitive device records the wave pattern made when light waves bounce off or scatter from the item. Gabor named this pattern a "hologram," derived from the Greek word "holos," meaning "whole," to emphasise that the whole wave pattern is caught in the photo.

Image of Dennis Gabor[4d]

Gabor's first attempt at making a hologram used a small, round, see-through piece with opaque writing[26]. The writing had the names of three prominent science figures—Huygens, Young, and Fresnel—who helped build the theoretical ideas behind Gabor's work. This breakthrough was first shared in the New York Times on September 15, 1948, in an article called "New Microscope Limns Molecula; Britons Impressed by Paper Combining Optical Principle With Electron Method." This key event significantly boosted Gabor's career, as his idea of "wavefront reconstruction" won praise from famous scientists like Charles Darwin (the grandson of the renowned evolution biologist), and Nobel prize winners in physics, Max Born and Lawrence Bragg[26].

Within the succeeding years, the wave-reconstruction technique received traction from scientists worldwide, with approximately fifty articles discussing Gabor's technique published from 1948 to 1955. The first Ph.D. thesis on holography was pursued in 1952 by Stanford's Hussein El-Sum. However, Gabor's theoretical and experimental conclusions had limited practical applications and usefulness during this period. Despite mathematically demonstrating the viability of holography using electron beams, his initial experiment involving a pinhole and a mercury lamp resulted in imprecise holograms[25]. Gabor faced disappointment as he couldn't develop a commercially practical system. The diffuse and unfocused nature of the available light sources at that time could have helped his goal of optimising the electron microscope resolution in that, almost ironically, Gabor's method often produced two unclear images instead of one[25]. This meant a low-quality in-line hologram was generated due to overlapping virtual and actual images. In other words, upon reconstructing the hologram, a virtual image formed at the original object's location, but its clarity was compromised by a coinciding false actual image. The images obtained were also small and blurry, leading to Gabor's growing frustration and unsuccessful attempts to convince colleagues to continue their investigations[25]. Reasons for the loss of interest include the continual failure to yield optimal results when the procedure was applied to an electron microscope and the troublesome flaws that the hologram's reconstruction process suffered from, as previously mentioned[25]. For a while, Gabor desperately tried experimenting with various optical configurations to minimise the impact of the conjugate picture, but by 1955, he had abandoned his holographic research.

Similarly, while Gordon Rogers of England was one of the most enthusiastic holographic researchers at the time, even he wrote this in 1956: "As far I am concerned, I am quite happy to let Diffraction Microscopy (linked to holography) die a natural death. I see relatively little future for it and am looking forward to doing something else."[25] It would seem that holography had reached a dead end and was no longer worth discussing, as seen when Gabor hardly brought up his work on "microscopy by wavefront reconstruction" upon his 1958 appointment to Chair of Applied Electron Physics at the Imperial College[25].

This is an instance where the laser can be seen as a miraculous invention. The true power of Gabor's work on 3D images came to be seen only after the 1960 invention of the laser. These lasers gave the robust and steady light needed for successful holograms, which Gabor couldn't achieve when using mercury lamps[25]. Indeed, two graduate students from the University of Michigan were able to create transparent holograms using Gabor's earlier ideas merely two years after Theodore Maiman made the first working laser[25]. Since then, the range of applications for holograms has become increasingly diverse, such as in medicine, the military, virtual reality, and security. This demonstrates an example where lasers, or photonics, have made a crucial impact on modern technology. This sudden eruption of interest in holography re-established Gabor's reputation, which was practically unknown in 1962. Still, he rose to win the 1971 Physics Nobel Prize "for his invention and development of the holographic method."[26]

To go into more detail, the two graduate students who invented practical holography were Emmett Leith and Juris Upatnieks. However, Leith's work on wavefront reconstruction extends before achieving this feat. After earning a master's degree in physics, Leith started working at Willow Run Laboratories' U-M Radar Laboratory in 1952[27]. Here, he helped in synthetic aperture radar (SAR). The Army wanted to build a high-quality radar imaging system, a goal they thought was impossible due to the massive antenna needed for high resolution[27]. However, this challenge could be overcome by synthesising the antenna's functionality, allowing a compact five-foot antenna to perform like one spanning a football field. The returns were captured as a single line on a photographic film for every transmitted pulse[27]. This process continued as the aircraft with the radar system flew and, similar to a hologram, produced a film strip that did not initially exhibit a recognizable image[27]. Processing of this radar data was then required.

In a concise period, Leith developed an entirely new theory of SAR based on physical optics. Leith acknowledged that "this new way of describing SAR in combination with optical processing is what today would be called a holographic viewpoint."[28] However, at the start, the scientific community recognized this theory more once its validity was demonstrated. Emmett's innovative idea for SAR processing, utilising wavefront reconstruction, was met with indifference from the SAR community and remained stagnant for approximately two years after its inception. It was only in 1957 that Willow Run Laboratories utilised these optical processing techniques to generate the earliest high-quality SAR images. Subsequently, the newly developed radar system was tested. The ninth flight yielded astonishing results despite eight flights yielding no discernible photos, which seemed to align with the sceptics' concerns and threatened the project's continuation. As Leith recounted, flight nine "produced some startling results. The terrain was beautifully mapped[27]. The Michigan SAR system became famous." This groundbreaking radar technology enabled detailed mapping of enemy terrain from a secure distance and had enhanced capabilities to penetrate fog and darkness beyond previous limitations. By 1959, Emmett's method of wavefront reconstruction had emerged as the predominant technique for optically processing SAR data[27].

As previously mentioned, the principle of holography involves using a photographic film to record a wavefield and subsequently reconstructing the wavefield later by illuminating the recorded hologram with a beam of light. While moderately successful, Gabor's attempts resulted in fuzzy images with a twin image that he couldn't resolve. Many scientists found this complication impossible, leading to a period of dormancy for holography following 1955[27].

This was until Leith, having worked on wavefront reconstruction for his SAR data processing efforts, developed an interest in holography[27]. In 1960, at the same lab, he contacted Juris Upatnieks, a new scientist there, to pique his interest in this novel study area. At first, Upatnieks wasn't impressed by what Leith had to say. But, after looking into Born and Wolf's "Principles of Optics," which talked about Gabor's work, he became intrigued. Despite Gabor's method's seemingly insurmountable challenges, they considered a new technique called carrier-frequency or off-axis holography[27]. What made this method stand out was using a second beam, known as a reference beam, which travelled around the object and hit the recording space from a side angle[27].

Image of Leith and Upatnieks at optical table [5]

Leith and Upatnieks came up with a method that presented terrific results. Leith said the hologram they produced was "indistinguishable from the original object itself, and the process required no more coherence of the light than the original Gabor process."[27] The coherence of light was the initial problem for Gabor. However, this new way, using a system with a unique frequency, cut down the coherence requirements by about 15% from what Gabor needed in his original approach[27]. This was a considerable step taken in early 1961, and it was published in the Journal of the Optical Society of America the following year[27]. In the fall of 1963, Leith and Juris showed the world display holography, although they called it lensless photography because no lens separated the object and the photo plate[27]. Even though they solved the big issue, more work and further enhancements were needed before people would recognize and adopt this method.

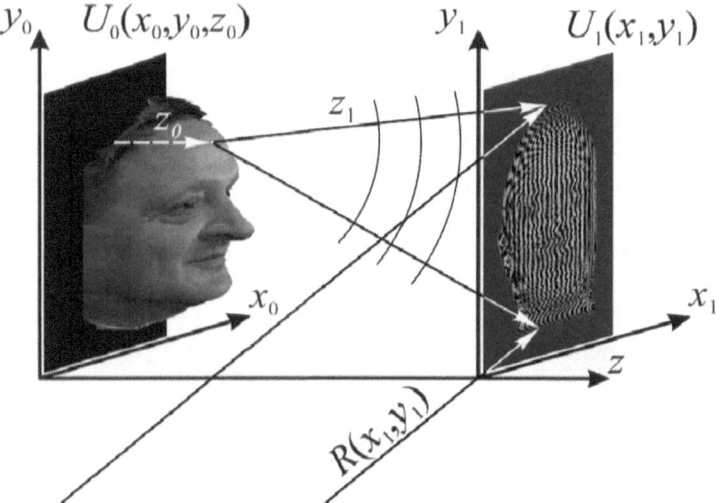

A diagram of the Hologram-recording-arrangement[6]

More specifically, researchers eagerly sought high-quality 3-dimensional display holograms, a feat yet to be achieved. Fortunately, the first helium-neon lasers for commercial use were introduced in about 1962, enabling Leith and Upatnieks to create more ambitious holograms[29]. They eventually produced holograms of considerable size (100 mm x 125 mm), visible to both eyes. At the Optical Society of America annual meeting in 1964, they astounded the optical community with an incredibly vivid hologram of a brass steam locomotive model[29]. The crystal-clear 3-dimensional representation of the train captivated the audience and resulted in world-widespread press coverage. Life Magazine covered Emmett and his groundbreaking hologram, while Scientific American also published articles about their work[27]. Impressively, these holograms possessed all the characteristics of real objects, including parallax, which refers to the apparent shift in position or direction of an object when viewed from different angles.

The exceptional realism and clarity of three-dimensional holograms captured the attention of a broad audience. This technology removed the idea that holography was a specialist or niche subject, resulting in worldwide interest in holography in general[27]. Holography had become almost extinct by 1960; however, starting in 1970, the situation changed dramatically, with many research groups from different regions of the world shifting their attention towards holography and coherent optics. Researchers have developed various strategies regarding advanced techniques, new types of materials used in making holographic images, and new applications, leading to the interdisciplinary nature this field has acquired. Industries like non-destructive testing, medical imaging, optical data storage, and entertainment have drawn researchers beyond the conventional optics field[27]. This innovation was truly revolutionary as it brought various communities together to work for a common purpose. In fact, in his Nobel Prize lecture, Gabor himself acknowledged the significant contributions made by Leith and Upatnieks. Besides the laser, he also attributed their success to the extensive theoretical groundwork of Emmett Leith, which began in 1955[27].

After his groundbreaking contributions, Leith continued expanding his work in this area. Coherent optics was the common principle uniting holography and SAR[27]. As a pioneer in the field, Leith made incredible breakthroughs in coherent optics applications and remained actively engaged in various imaging-related areas throughout his career. He continued to find research partnerships to improve these technologies and formed connections with many individuals seeking his extensive expertise to guide their research endeavours[27].

In 1987, Dr. David Dilworth began collaborating with Leith during his doctoral studies with regular lab meetings. Needless to say, their joint research primarily revolved around coherent optics but specifically focused on enhancing imaging techniques, especially in biomedical imaging. Dilworth and Leith hold a patent associated with identifying breast cancer at its earliest stages[27]. This patent pertains to the concept of photon migration, which refers to capturing images through mainly scattering media such as biological tissue[27]. Over time, photon migration has evolved into a significant field within optics. Once again, Leith's involvement in this area dates back to the late 1980s, when he utilised holographic methods to achieve the seemingly impossible—rendering biological tissue crystal clear like transparent glass. By accomplishing this feat, the intention was to enable visual inspection of living tissue, facilitating the identification of abnormal growths, including malignant tumours[27].

Dr. Brian Athey, a psychiatry professor and director of the University of Michigan's Visible Human Project, was another expert Leith has worked with[27]. Their joint projects began in the early 1990s. Within the Visible Human Project, Athey aimed to present a three-dimensional image to viewers that removed the need for 3D glasses. This project aimed to replace traditional anatomical dissections with virtual alternatives. This shift began in experimental educational environments and computer-based surgical training, hoping it would become common in the next 10-20 years[27]. As a side note, besides his work in the lab and his passion for optics, Leith found joy in teaching. In the 1960s, he helped start an optics program and developed a new passion for teaching students. One of his students said he was "always appreciative of the exciting phenomenon of light."[27] Throughout his life, Leith played such a significant role in the advancements of this field that in 2006, the Optical Society of America established The Emmett N. Leith Medal in his honour, honouring his revolutionary work in the field of optical information processing[27].

Fibre optic cables

Focusing on the 1970s, using fibre optic cables in telecommunications represented a significant turning point. However, the underlying technology behind it has a much longer history. Impressively, as early as 27 B.C., people began drawing glass into fibres during the Roman era. In the 1790s, the French Chappe brothers invented the first optical telegraph, consisting of towers equipped with lights that operators could employ to relay messages[30]. Advancing further, physicists Jacques Babinet and Daniel Collodon proved in the 1840s that light can be guided along water jets used in fountain displays[30]. Using this concept as a foundation, British physicist John Tyndall illustrated in 1854 that light signals could be bent by directing them through curved streams of water[30]. These discoveries laid the scientific groundwork for modern fibre optic technology. Another inventor, Alexander Graham Bell, is renowned for inventing the telephone. Less known about him is that in 1880 he patented an optical telephone system called the photophone, which used a beam of light to transmit sound. Around this period, William Walter developed a system of "light pipes."[30] These pipes incorporated reflective coatings to illuminate houses by using pipes to direct light from an electric lamp placed in the basement throughout the house[30]. Unfortunately, this method proved ineffective and was overshadowed by the following success of Thomas Edison's incandescent light bulb. It would seem that these early innovations did not attain much success. However, although these early innovations did not achieve widespread recognition, that is not to say they were invented in vain. Actually, they served as crucial initial steps in the advancements that paved the way for our modern technology, leveraging light as an efficient transmission channel.

The advancement of fibre optic technology opened up new possibilities in various industries outside of telecommunications. One particular field would be medicine and surgery. In 1888, physicians Roth and Reuss from Vienna utilised bent glass rods to provide illumination within body cavities. A decade later, American inventor David Smith sought a patent for his dental illuminator incorporating a curved glass rod[30]. These innovative applications demonstrated the potential of fibre optic technology in advancing medical and dental practices. Surgeries became more successful, and this success of fibre optics technology slowly made its way to innovations in more industries. Moving into the 20th century, in the 1920s, the notion of transmitting images for television using an array of transparent rods and facsimiles was patented by John Logie Baird and Clarence W. Hansell, respectively. Having said that, one of the earliest investigations of television technology was conducted in the late 19th century when French engineer Henry Saint-Rene created a framework of bent glass rods to direct light pictures[30]. However, Heinrich Lamm achieved the milestone of transmitting an image via an optical fibre bundle in 1930[30]. The image he successfully transmitted depicted a light bulb filament. His goal was to be able to observe body sections that were not easily accessible, but the change in the power dynamics during World War II meant he had to abandon his research. Due to Hansell's British Patent, his attempt to apply for a patent was unsuccessful[30]. Hansell's patent and Baird's patent also prompted the denial of Danish physicist Holger Moeller's patent in 1951, where he suggested the idea of coating low-index materials around glass or plastic fibres for fibre-optic imaging. Later, Abraham Van Heel, known as the father of technical optics in the Netherlands, produced a cladded fibre system that significantly decreased signal interference and problems that arose between fibres, like crosstalk[30].

New technological possibilities were explored after the laser was introduced in the 1960s as a more efficient and precise way of transmitting light than standard light.

American Optical's Elias Snitzer published a theory in 1961 to describe single-mode fibres whose core would be so tiny that it would only require one waveguide mode to transport light, meaning there is only one way the electromagnetic field exists within the waveguide[30]. Snitzer successfully showcased a laser-guided through a narrow glass fibre, which proved suitable for medical applications. However, when it came to communication purposes, the issue of light loss became a significant obstacle, making long-distance transmission impractical. Light loss, also called attenuation, is an inherent by-product of transmitting signals over extended distances. Fortunately, this problem was quickly addressed only after a few years when Charles Kao put forth a theoretical demonstration of how eliminating glass fibre impurities could minimise light loss[30]. Kao's groundbreaking work in this area earned him the 2009 Nobel Prize. Practical telecommunications using optical fibres was now possible by combining laser technology with Kao's solution.

In 1970, scientists at Corning Glass Works (presently known as Corning Incorporated) in the U.S. successfully created a low-loss optical fibre, who reported on their prototype silica glass optical fibre with a transmission loss of 20 dB/km[30]. Over the years, fibres have continued to be created with increasingly low losses. In 1973, a 2 dB/km transmission loss was obtained by Bell Laboratories by developing a modified chemical vapour deposition process that creates glass layers in a silica tube by heating oxygen and chemical vapours, more specifically, raw glass ingredients in the gas phase. Low-loss optical fibre can be made by mass-producing these glass layers[30]. This method spread to other countries and remains the standard for fibre-optic cable manufacturing. Soon after, telephone companies quickly replaced their communications infrastructure, essentially using fibres towards the end of the 1970s and early in the 1980s[30]. Sprint was the first ever U.S. telecom firm to build a nationwide network that was 100% digital and composed of fibre-optic cables in the mid-1980s. Then, in 1988, the first-ever transatlantic telephone cable using optical fibre began service, connecting France, the U.K., and the U.S. The technology was further improved in 1991 when a new, all-optic fibre system was created to transport 100 times as much data as cable[30]. The first all-optic fibre cable was installed beneath the Pacific Ocean in 1996. The following year, in 1997, the Fibre Optic Link Around the Globe (FLAG) emerged as the world's longest single-cable network[30], with an impressive length of 28,000km, connecting the UK, India, Japan, and many other places. Over the next decade, fibre-optic cables carried nearly 80% of the world's long-distance traffic data; nowadays, billions of kilometres of fibre-optic cable have been deployed around the globe[30].

Photovoltaic devices

Besides devices like photodetectors, which detect photons or optical signals and translate them into electrical signals via internal photoelectric effects, and light-emitting diodes (LEDs) or laser diodes, which convert electrical or light energy into coherent beams of light, there are other types of photonic devices. Photovoltaic devices are solar panels that directly convert energy from sunlight into electrical power using photovoltaic effects. At the young age of 19 in 1839, Edmond Becquerel, a French scientist, discovered the photovoltaic effect. He realised that electrons in a conduction band could flow freely through a material when excited, which created a current. However, his discoveries weren't widely recognized until Einstein wrote a paper about the power of solar, for which he was eventually awarded the Nobel Prize in 1922.

Even as early as the 7th Century B.C., humans have harnessed power from the sun for various purposes, with early uses primarily for creating fire. Moving forward, Augustin Mouchot, a French maths teacher, produced several solar-powered machines in 1861 because he was worried that the world's coal reserves would ultimately run out, starting with a solar oven for cooking and a solar water heater for baths. It was then, in 1866, that he developed the first solar-powered steam engine[31]. It worked by focusing the sun's rays onto a water-filled tube through a curved, shiny metal trough. This created steam, which powered an engine. At the 1878 Paris World's Fair, Mouchot displayed his greatest solar engine with demonstrations such as a refrigeration system that created blocks of ice using heat[31]. Needless to say, the sight of solar-powered ice stunned the audience, creating a sensation. Alas, the 1878 show also saw a surge in the popularity of the internal combustion engine right when oil and coal were falling in price. As a result, fascination with solar energy and engines quickly faded[31]. However, as time passed, people began to realise the effects of non-renewable energy, such as coal and the possibility of their depletion. As a result, the search for renewable energy sources once again resurfaced and thus began the modern renaissance of solar power.

Despite that, the first solar panel was actually invented by Charles Fritts in 1883, where he coated a thin coating of selenium with an incredibly thin gold layer. Although the resultant cells only had a 1% conversion electrical efficiency, far from enough to power electrical equipment, they proved that a solid material could change light into electricity without heat or moving parts[32], thus leading to the beginning of a solar energy production movement.

Because of the initial poor output, today's solar cells are composed of silicon rather than selenium[33]. Gerald Pearson, Calvin Fuller, and Daryl Chapin of Bell Labs were credited with this innovation in 1954. Their solar cells produced an efficiency of only 4%, but this was the first time solar technology could sustain an electrical device for multiple hours each day[33]. This breakthrough led the U.S. government to pour more money into solar cell technology. In 1973, the University of Delaware achieved a significant milestone by constructing the first-ever solar building, known as "Solar One." This innovative system is operated by using a hybrid combination of solar photovoltaic and solar thermal power sources. Additionally, it was the very first example of building-integrated photovoltaics[33]. Instead of using solar panels, the rooftop was integrated with photovoltaic materials, which is comparable to the Tesla new roof product's design[33]. From 1957 to 1960, the efficiency of photovoltaic devices increased by 14% from 8% with the breakthroughs from Hoffman Electronics[33]. In 2022, the world record for solar cell efficiency is 47.6%, and further developments are only to be expected with the growing demand for renewable energy. In fact, in 2022, the solar panel industry was valued at at least over a hundred billion USD. The falling costs of solar energy are one reason for its growing popularity. The significant drops in solar panel costs over the previous few decades have led to a rise in customer demand in the U.S. in early 2016 that required over a million installations[33]. Solar panels cost 300 USD per watt in 1956, and by 1975, that value had fallen to a little over $100 USD per watt. Since 1980, the cost of solar panels has decreased annually by at least 10%, allowing for solar panels prices to go as low as 0.50 USD per watt[33].

Lighting

Moving on to something perhaps simpler, lighting technology. Although the invention of the laser brought immense progress to the photonics field, the improvement of general lighting technology is just as important. One of the first forms of man-made lighting was torches, where there would be a shell packed with combustible material, scattered with animal fat, and ignited. Next, there were candles, with the Ancient Egyptians frequently credited with being the earliest to use them, although these did not have wicks like a true candle[34]. Wicked candles are often credited to the Romans, but many different civilizations around the world developed their own kind of candles separately, one example demonstrating the universal motive of improving the ways we use light. Advancing to closer times, late in the 1790s and early in the 1800s, gas lamps and electric lamps were invented until finally, in 1879, the electric light bulb was invented. Thomas Edison, an American inventor, set out to construct an electric light bulb with a long lifespan capable of rivalling gas lighting. On October 22, 1879, he successfully created his first incandescent light prototype, which burned for 13.5 hours. This was greatly improved only after a few months when he found a carbonised bamboo filament that could last up to 1200 hours[35]. This was the game-changing development he had been searching for; it was the lighting technology that was needed to make electricity the norm for powering both indoor and outdoor lighting[35].

Many other lighting advancements were made during the 20th century, such as the fluorescent bulb, which replaced the filaments in a bulb with inert gases. However, there was one invention that was clearly distinct from its predecessors. Previously, light bulbs operated by supplying filaments or gases with an electrical current to heat them and produce light. Conversely, LED lighting is a solid-state light, a kind of lighting that emits light through semiconductor diodes[36], which convert electrical current and release energy as photons using a semiconductor in order to generate light. This use of electronic technology to convert and emit optical radiation makes it an optoelectronic device. In 1961, an infra-red LED was invented by Robert Biard and Gary Pittman as a by-product of the pursuit of creating a semiconductor laser in the visible range[37]. But since this light was beyond the range of human visibility, it served no practical useful purpose. The 1960s saw the invention of the first visible LED by scientist Nick Holonyak, who was working at General Electric and referred to it as "the magic one."[37] This was a red LED. In recognition of his groundbreaking work, Holonyak is widely regarded as the "Father of the Light-Emitting Diode."[37] Applications of this invention can be seen as early as 1964 when IBM started to implement LEDs for the first time on a circuit board in an early computer and in 1968 when Hewlett Packard (H.P.) started incorporating LEDs into its calculators[37].

Other colours of LED were gradually developed using different chemicals, but the last colour, blue, had proven to be a challenge to produce. Blue LEDs required chemicals, including carefully created Gallium Nitride crystals, that scientists weren't yet able to make in the lab. The invention of the blue LED was so groundbreaking that, in fact, the three pioneers of the blue LED, Isamu Akasaki, Hiroshi Amano, and Shuji Nakamura[38], were awarded a Nobel prize with the statement: "Incandescent light bulbs lit the 20th century; the 21st century will be lit by LED lamps."[39] They completed the RGB spectrum and allowed for the production of visible full-colour LED screens and white LEDs. The arrival of white LED light has enabled companies to develop light bulbs, computer screens, and smartphones[40] that possess improved longevity and energy efficiency, outperforming any previously invented bulb.

However, although LEDs have actually been around since the 1960s, they did not gain traction until the mid-2000s. Since light-emitting diodes only gained prominence late in the twenty-first century, some people believe that they are a relatively new invention. The cause of this late upsurge in popularity simply had to do with money. In 1963, General Electrics began offering red LEDs, which were initially priced at $200 for each lightbulb[41]. This was seen as exorbitant since people were accustomed to paying between $5 and $20 for fluorescent and incandescent bulbs, so it was initially a challenge to enter the market[41]. By the mid-1970s, even though Fairchild Optoelectronics created an LED that could be manufactured at the cost of 5 cents, the limited production batches meant that the price to customers stayed extremely high[41]. As large companies like IBM and H.P. began utilising LEDs over time, the technology started to market itself [41]. Production levels increased in tandem with demand. Manufacturers were able to use mass production to capitalise on cost savings through economies of scale, which resulted in a sharp decline in cost per unit.

This invention of the LED revolutionised how we light up our environment. They cost less to operate and emit less heat than incandescent bulbs, and some LEDs can operate at roughly 10% of the power required for the standard light bulb and last up to 40 times longer. They are used everywhere, from illumination systems to display screens like televisions. The iconic New Year's Eve ball, which drops in Times Square annually, comprises a remarkable 32,256 individual LED lights. Furthermore, advancements in LED technology have given rise to flexible LEDs, which are suitable for use in devices like foldable smartphones and wearable electronics. Notably, researchers at Zhejiang University in China have achieved a significant breakthrough by developing flexible LED screens that are also transparent[42]. These innovative screens use silver nanowires, which enhance the existing manufacturing processes, showcasing the potential of silver nanowires to replace conventional displays. Before attempting to successfully integrate into a commercial product, the group is committing to further research on how best to protect and maintain the circuitry[42]. They anticipate that these nanowires could also open up a wide range of possibilities for built-in displays in windows and buildings[42].

Metamaterials

Next, moving on to metamaterials. "Meta" originates from the Greek word "beyond," μετά. In this case, it refers to how metamaterials are materials artificially engineered to exhibit characteristics that are rarely observed or even not available in nature. They are made of subwavelength building blocks made of composite materials, a combination of two materials with different physical and chemical properties, which allow for extreme control over optical fields. Following World War II, much of the research related to metamaterials was on the use of artificial dielectrics in the microwave regime, especially for antenna beam shaping, as it was proposed as a low-cost and lightweight "tool". For relevant parts of the electromagnetic spectrum, investigation on artificial dielectrics or artificial insulating materials is still in progress.

In the 20th century, many of the studies on metamaterials were mostly theoretical due to the difficulty of fabricating these structures at optical scales. For example, negative-index materials were first described theoretically by Victor Veselago in 1967[43]. These would be materials with negative refractive indices and are part of left-handed materials, which are materials with opposite optical effects compared to natural optical materials like glass, i.e. negative permittivity, permeability and refractive index. In the area of electromagnetism, permittivity and permeability are distinct measurements. Permittivity quantifies a material's capacity to store energy internally, while permeability gauges a material's capability to facilitate the creation of a magnetic field within itself. These materials are now a crucial part of metamaterials. However, it took three decades for the first useful metamaterial based on Veselago's ideas to be constructed[43].

The progress can be seen in the 1990s when Pendry et al. created a structure that could regulate the magnetic interactions of light radiation, even if it was limited to microwave frequencies[43]. The material magnetic parameters were extended into the negative by this split ring arrangement that repeated progressively. This "magnetic" structure, which is lattice, as in periodic, was built using non-magnetic elements. In this experiment, it is demonstrated that in order to create the first metamaterials, simultaneous negative values for permeability and permittivity were necessary. These were the initial steps toward proving the viability of Veselago's 1967 idea. The creation of the initial metamaterials was made possible by the developments in fabrication and computing capabilities during the 1990s[43]. Testing of the "new" metamaterial for the phenomena mentioned by Veselago started, but initially, it was restricted to the microwave range[44].

It wasn't until right at the turn of the millennium, in 2000, that researchers at UCSD developed and presented metamaterials with peculiar physical characteristics not found in any other substance found in nature[43]. Despite adhering to the laws of physics, these materials behave in a unique way from typical materials. Essentially, several of the physical characteristics dictating the behaviour of typical optical materials could be reversed by these materials. Among such unusual qualities is that this was the first time we were capable of reversing Snell's law of refraction[43]. The UCSD team's May 2000 display marked the material's public debut.

David Smith was among the researchers at UCSD who contributed to the creation of the metamaterials. In his account, he states that metamaterials, a type of artificial material, were just a hobby to try and comprehend a completely unrelated phenomenon. Negative index materials and the "magical properties" of metamaterials were nothing he ever imagined would exist[45]. Plasmonic materials are materials with negative permittivity, and the primary obstacle in developing a microwave plasmonic material was the lack of known materials possessing the necessary characteristics. This changed with Pendry's discovery, and while looking through the large collection of literature on plasmons, a 1996 study by Pendry et al., featured in Physical Review Letters, caught Smith's attention. Here, he learnt of the suggested artificial material composed of wires, which could exhibit plasmonic behaviour at any frequency. Unfortunately, though, none of Smith's theoretical colleagues at UCSD understood Pendry's paper when he showed it to them[45]. Moreover, controversy surrounded Pendry's theory at the time, with many objections to Pendry's approach and results from many scientists and theorists.

In 1998, when PIERS (Progress in Electromagnetic Research Symposium) was held, Smith was invited to attend. By surprise, that conference proved to be incredibly pivotal and serendipitous. In Smith's session, many people talked about Pendry's wire medium. Some groups actually succeeded in confirming Pendry's predictions through comprehensive numerical simulations. Though no experiments were conducted, increasing evidence supported the notion[45]. Still, extremely small wires were needed. Secondly, it turned out that also attending that particular conference session was Eli Yablonovitch, another famous name to be discussed later. Smith had known Yablonovitch for years, and when they started comparing notes, it came up that he was also highly interested in wire architectures and the potential existence of microwave plasmons[45]. Then and there, Yablonovitch invited Smith to talk at a conference on photonic crystals in a couple of months. Smith was slightly sceptical as he did not know of the microwave plasmons. Still, Yablonovitch was very persuasive and convinced Smith to join by agreeing to allow for a failsafe topic[45].

As time passed, Smith struggled to understand Pendry's paper, and more than once, he called Yablonovitch to request to switch his topic as agreed, but each time, his request was shut down[45]. Therefore, with just three weeks remaining until the conference, Smith was experiencing extreme panic. However, this stress and Yablonovitch's enthusiasm pushed him to sit down and think about the paper seriously. Pendry presented a very sophisticated, refined theoretical hypothesis for a collection of tiny wires' effective medium response. Furthermore, the findings had been presented using terminology that would be easily understood by a condensed matter theorist, which Smith was not. However, Smith thought perhaps there was another way to reach his formulas, which ended up being really simple. The moment he began to consider a different derivation, everything fell into place almost instantly[45].

At last, he grasped the content of Pendry's paper and conceived of a method for creating a microwave plasmonic material. However, to make a conference poster, there were a lot of tasks to complete. One initial issue was that this project was completely unrelated to the optical nanoparticle microscopy that Smith was initially studying and undoubtedly did not have a connection to the business Seashell had started, which tried to commercialise the metal nanoparticles so they could be used in biomedical analyses as incredibly bright labels, so there wasn't much of a case to be made for formally researching this topic with lab resources[45]. However, in their university research group, the postdocs and students were all amiable, happy, and excited to work on new projects, even if it required extra work and projects that needed to be completed outside regular work hours. They had to spend their nights and weekends working on this. Students and researchers could help with simulations, measurements, and samples, for example. They all enthusiastically volunteered to assist Smith when he appealed for their help in bringing the concept of plasmonic mediums to reality. Therefore, within two weeks, they formulated the conceptual framework for the loop-wire medium, ran the simulations, fabricated the material, and created the plasmonic nanoparticle equivalent in the microwave[45]. Shockingly, everything went according to plan. Yablonovitch's friendly pressure had proven to have been an effective catalyst[45].

This was not the end, though. At the workshop where the poster was to be presented, Smith and Schultz met John Pendry, who agreed to look at their poster. In just a couple minutes, he affirmed that they possessed a comparable method for reaching a microwave plasmonic medium and found the loop wires they had created to be really fascinating. Shortly after, during the workshop, Pendry presented his initial lecture on the split-ring material he had discovered[45]. Pendry spoke of "split-ring resonators," another common component in photonics. Even though these structures lacked naturally magnetic materials, Pendry anticipated they would possess a magnetic resonance with controllable permeability areas, and therefore, he was attempting to manufacture artificial magnetism[45].

Since the workshop was mainly focused on photonic crystals, not many were interested in Pendry's topic, except Smith. So right after, he got Pendry's permission to start using his unpublished manuscript as a model to demonstrate artificial magnetism in the laboratory, as the university group he worked with was also enthusiastic about man-made electromagnetic structures now that everything appeared to be functioning so swiftly and efficiently[45]. Pendry did caution them that it would be probable that his team would beat them to publishing as he was collaborating with others to confirm the assumptions and was already quite ahead in progress[45]. This did not matter much since Smith did this as a hobby; they had no intention of competing with anyone. It was difficult as they had to figure out the details of how to construct a functioning resonator, such as the techniques to produce them and the properties of the produced materials. Eventually, after simulations, they gave the designs to Willie Padilla, who was in charge of making samples. Within a day, he succeeded in making working artificial magnetic materials[45]. They were so impressive that they even defeated Mike Wiltshire, who had attempted the same strategy and got extremely close. Still, he could not establish their qualities conclusively because he was not conducting simulations[45].

Now, there is extremely solid and convincing proof that the array of conducting rings acted like magnetic material thanks to this experiment and the simulations. They had actually developed an excellent set of interpretations and experimentation for an unsponsored "hobby" endeavour[45]. Although it had taken a lot of work to demonstrate the rings' magnetic response, the outcome was satisfactory and unquestionably sufficient to warrant writing a manuscript. Smith felt that the physics from this study was distinctive enough to be submitted to Physical Review Letters (PRL), the most esteemed physics journal[45]. Together with Willie, he completed the manuscript, which they submitted in December 1999. Their manuscript was turned down right away, which Smith did not find surprising as he thought that since the experiment lacked anything genuinely groundbreaking from a physics perspective, it seemed to relate more to engineering than physics[45].

However, after doing some research, Smith found Veselago's paper. With this discovery, Smith and his team realised that their work actually demonstrated a prediction from 3 decades ago. They had created the first harmful index material. This changed everything. Schultz and Smith rewrote the abstract to emphasise a focus on the discovery of a material with a negative index instead of just showing proof for split rings being artificial magnetic materials by paraphrasing Veselago's predictions[45]. The phrase "metamaterial" has been circulating around in the physics field. Since they were sure their structure might also be regarded as such material, they included the term in their abstract and discussion[45]. A couple of months after resubmission, the manuscript was finally accepted for publication.

For them, the approval of the PRL was a huge victory, and Smith was pleased that they had managed to produce some compelling physics. Still, he didn't think the harmful index material would benefit real-life applications. Doubting that much future research would be carried out. He believed this was the end of his journey with synthetic materials as nobody would be willing to provide funding for this type of work, or so he thought, and was returning to the research of optical plasmons[45]. Evidently, he had greatly underestimated his discovery.

Although the metamaterials discussed controlled microwaves, they also have their applications in photonics. One particular application is in superlenses, mentioned in the introduction chapter. Utilising the unusual properties of metamaterials, such as the negative refractive index, scientists are working towards the ideal of infinite resolution, the ability to zoom in as far as possible on an image while maintaining an appropriate level of image resolution. This means superlenses benefit fields like biomedical imaging, microscopy, and nano-optics. Superlens construction was considered impossible at one point. In 2000, Pendry claimed that using left-handed materials would do the job. However, a practical demonstration of such a lens took longer to develop because, as mentioned earlier, it is challenging to fabricate metamaterials with negative permittivity and permeability properties[46]. Pendry went on to propose more theories for approximate superlens to be realised. Pendry's concept was first experimentally shown at RF/microwave frequencies in 2003. Two other groups then independently proved Pendry's lens in the UV range in 2005. A functioning superlens at microwave wavelengths was constructed in 2004 by electrical engineers Anthony Grbic and George Eleftheriades, and also in 2005, an optical superlens with resolution three times better than the conventional diffraction limit was experimentally demonstrated by Xiang Zhang and associates.

Photonic crystals

Eli Yablonovitch, introduced earlier, is credited with developing the idea of the photonic bandgap, and it was him who coined the term "Photonic Crystal."[47] Photonic crystals are optical media containing a periodic nanostructure, structures made of high and low refractive indices regions that infinitely or finitely and periodically repeat in one to three dimensions[48]. Photonic crystals have some remarkable properties, such as a complete photonic band gap. According to their wavelength, light waves may move through the crystals or propagation could be disallowed. Bands are the wavelength ranges which propagate. Photonic band gaps are blocked wavelength bands[48]. The photonic bandgap is an essential concept because it has the capacity to contain and regulate electromagnetic radiation in all three spatial directions.

While photonic crystals have been investigated since 1887 in one way or another, the phrase "photonic crystal" was not used until 1987, more than 100 years later, following the release of two seminal studies by Sajeev John and Eli Yablonovitch on the topic. After 1987, research papers about photonic crystals soared exponentially in numbers[49]. Nevertheless, because these structures are challenging to construct at optical wavelengths, much like metamaterials, early research was either theoretical or conducted in the microwave spectrum, where photonic crystals may be created on a more accessible centimetre scale[48]. Where it is possible to make photonic crystals on a centimetre scale that is easier to access, an important thing to note is that while photonic crystals and photonic metamaterials have similarities, they are different in the principles they use[48]. A distinction is that the resulting exceptional properties of metamaterials are not explained with photonic band structures but with unusual refractive index values. Also, unlike metamaterial structures, photonic crystal structures have been observed in nature. For example, these structures occur naturally in minerals like opal and living organisms like butterfly wings.

In 1991, Yablonovitch presented the first photonic bandgap in three dimensions in the range of the microwave regime[48]. By drilling a transparent material with an array of holes drilled in each layer forming an inverse diamond structure, Yablonovitch created the first 3D photonic crystal with a complete photonic bandgap, which blocks light from passing through at specific wavelengths[49]. "Yablonovite" is the name given to this crystal, and the discipline of photonic band engineering was established because of the impact of his work. Nowadays, photonic crystals are a crucial component of numerous technologies, such as optical fibres for telecommunications or light-emitting diodes, to increase efficiency. Additionally, in cancer surgery, these crystals assist in focusing infrared laser beams to eliminate tumours[50]. Yablonovitch co-founded Luxtera Inc., one of the originators and world leaders of Silicon Photonics, now acquired by Cisco Systems. Each Luxtera Silicon Photonics chip has a 2D photonic crystal[51]. These chips are used by billions of people worldwide, and millions are found in large data centres[51].

Pavel Cheben demonstrated subwavelength grating (SWG) waveguides, a type of photonic crystal waveguide. Waveguides are another essential component in photonics technology, explained later. The SWG waveguide functions away from the bandgap in the subwavelength range[48]. It reduces the effects of wave interference and permits direct control of the waveguide properties through nanoscale engineering of the resulting metamaterial. This provided "a missing degree of freedom in photonics" and addressed a significant limitation of silicon photonics: its narrow range of accessible materials was insufficient to perform intricate optical on-chip tasks[48]. He introduced a novel area of science that combines metamaterial research and integrated photonics. Researchers worldwide are exploring more methods for integrating photonic crystal slabs into integrated circuits, such as to enhance the optical processing of communications[48].

Optogenetics

A slightly more recent development in photonics would be optogenetics. In 2006, Karl Deisseroth et al. first introduced the term "optogenetics" to broadly refer to a method for controlling and observing the biological activities of cells and tissues that have been altered to express photosensitive or light-sensitive proteins using genetic engineering and optical technology[52]. More specifically, it is used especially in neuroscience as a biological technique to control the activity of neurons or other cell types with light[53]. The capacity to observe and analyse living, moving animals is one of optogenetics' biggest advantages. Historically, Francis Crick proposed in 1979 that light might be used to rapidly provide spatiotemporal control for targeting particular neurons because light can be turned on and off [54]. This idea led to the development of optogenetics as a field in neuroscience. In his 1979 article, Crick explains that controlling one type of brain cell while leaving the others unaltered presents the biggest problem for neuroscientists[55]. This is unobtainable with electrical stimulation since it excites all nearby cells equally. In contrast to the millisecond timescale of brain activity, drugs act too slowly[55]. However, at that point, the method to use these photosensitive proteins in neuroscience to increase brain cell sensitivity to light was unknown to neuroscientists[56].

In 2005, the groundbreaking paper "Millisecond-timescale, genetically targeted optical control of neural activity"[57] written by Karl Deisseroth et al. was published in the journal Nature[58]. The significance of this research paper was that it was the first to thoroughly explain optogenetics as a method of making neurons light-sensitive so they could then be stimulated by light. The work demonstrated how researchers may activate or inhibit particular brain neurons and examine the results by fusing genetic advances with optical technology.

As a matter of fact, Deisseroth is sometimes known as the father of optogenetics. However, there is another name that is not as known — Zhuo-Hua Pan. Pan's motivation was to find a cure for blindness. Blindness is a consequence of the loss of rods and cones in the eyes, responsible for light and colour detection. Pan's idea at the start of the 2000s was that blind people could see again if a light-sensitive protein was injected into their eyes to cause other cells in the eye to become light-sensitive. The concept behind optogenetics originated with introducing a protein into neurons that transforms light into electrical activity[59]. In this manner, researchers may remotely trigger the neurons with light and manipulate brain circuits. Previous attempts to add light sensitivity to neurons had been made by others, but those methods had failed because they were missing the proper light-sensitive protein.[59]

All of that changed in 2003 with the publication of the first explanation of channelrhodopsin's molecular structure[59]. Green algae produce this protein called channelrhodopsin, which reacts to light by releasing ions into cells, helping the algae seek sunlight. By February 2004, Pan had grown ganglion cells, the eye neurons that had direct connections with the brain, contained in a dish and was experimenting with channelrhodopsin in these neurons. The presence of light caused them to become electrically active. Pan, overflowing with excitement, applied for the National Institutes of Health funding. With the remark that his study was "quite an unprecedented, highly innovative proposal, bordering on the unknown," the NIH granted him a $300,000 grant[59]. Pan had no idea at the time that he was competing against research teams around the nation and the world to implant channelrhodopsin into neurons, teams that included Deisseroth and Boyden at Stanford, among others.

In March 2004, Deisseroth contacted Georg Nagel, the paper's author, and requested to collaborate. This way, Boyden could obtain the channelrhodopsin DNA and test it in neurons. Then, it was in August 2004 that Boyden observed the channelrhodopsin's electrical activity in a brain neuron illuminated in a dish.[59]

Pan had also experimented with this approach with retina neurons six months prior. However, he was scooped. He noted that scientists build upon each other's work, sometimes collaborating and other times working concurrently and scrambling over each other's shoulders[59]. To this, he said, "It's funny to think about how science regards when something is proven." He also added that when scientists work on the same topic, "There's both intentional and unintentional teamwork."[59]

Pan worked on getting the protein channelrhodopsin into a functional eye throughout the summer of 2004. He ultimately decided to use a virus that could secrete the channelrhodopsin DNA into the eye's cells through infection[59]. To track the location of the channelrhodopsin, his colleague Alexander Dizhoor modified the DNA of the channelrhodopsin to include the gene encoding a protein that fluoresces green in response to blue light[59]. This falls under the fluorescent techniques discussed later in the book. Pan administered the virus to his first rat in July 2004[59]. He examined the retinas to determine whether it had been effective around five weeks later. What he observed was "a sea of green" — the ganglion cell membranes with a number in the thousands contained the green protein connected to channelrhodopsin, and there was a spike in electrical activity from that cell when he inserted an electrode in it and switched on a bulb. The application of channelrhodopsin was successful. Albeit it was only a first step, it was a revolutionary one, suggesting the immense and exciting potential Pan's technique had in eventually being able to give blind people their sight back.

On 25 November 2004, Pan and Dizhoor submitted an article detailing their study to Nature based on this novel discovery[59]. However, the Nature editors recommended sending it to Nature Neuroscience, a more focused journal, which rejected it[59]. When Pan submitted the work for evaluation to the Journal of Neuroscience early the following year, he was turned down once more. Pan, discouraged, went to work editing his paper and travelled to Fort Lauderdale for the Association for Research in Vision and Ophthalmology conference in May 2005, where he gave a talk about his research on channelrhodopsin in neurons. His greatest contribution to the history of invention would be that one fifteen-minute presentation[59].

The reason that matters is because of what happened next: in August 2005, just a couple of months later, a publication on the application of channelrhodopsin to increase light sensitivity in neurons was published in Nature Neuroscience. The paper was written by none other than Edward Boyden and Karl Deisseroth[59]. After Pan received the news about the paper via an email from a coworker, he said, "I felt terrible. I felt terrible," and "We didn't feel very lucky."[59]

On the one hand, in comparison to Pan's paper, Deisseroth and Boyden's had a few slight differences. They merely showed that they could control the activity of neurons in a dish using channelrhodopsin; Pan had held off on publishing until he could successfully control neuronal activity in a living animal. Deisseroth and Boyden also demonstrated extraordinarily accurate time control by activating the light for a mere millisecond. However, they achieved the same technical feat: they had effectively used channelrhodopsin to induce light-induced responses in neurons cultured in a dish[59].

Even the Stanford paper took a while to gain attention, but once it started to, it took off. Boyden at MIT and Deisseroth at Stanford each received considerable funding and gifted pupils for their laboratories following the work's success, which launched both their careers. Furthermore, after Deisseroth The New York Times covered Deisseroth's optogenetics accomplishments in 2007, the study paper's citations skyrocketed[59].

On the other hand, most people dismissed Pan's research when he was able to publish his research in Neuron at last in April 2006; most people just shrugged it off. For example, neuroscientist Richard Kramer of UC Berkeley, who was also investigating vision, described it as unimpressive as it seemed to reiterate the paper on putting channelrhodopsin in neurons, only this time with applications in the retina[59].

The answer as to why Pan's paper wasn't published first may never be known for sure. However, Pan did write to the editor of Nature Neuroscience after Boyden's study was published, asking for an explanation as to how Boyden's paper got accepted when they rejected him. The response he received was that despite the similarities between the publications, Boyden et al. portrayed theirs as a novel technology. Pan's work appeared to be limited, mainly focusing on his scientific results on using channelrhodopsin for vision restoration, whereas Boyden's work took a broader approach, seeing it as a tool for neurology in general[59]. It's also possible that his attendance at Wayne State University prevented Pan from having the resources to publish a high-profile paper. In addition to the direct expenses associated with doing high-calibre research, junior professors at prestigious universities are typically mentored by experienced researchers who offer feedback and assist them in elevating their work to the next level. Pan acknowledges that this fact might have made him less competitive than scientists from elite universities like Stanford or MIT[59].

It's interesting to note that Pan's May 2005 talk jeopardised the Boyden-Deisseroth patent for a while because the US patent office denied it. After all, Pan's abstract had already been public over a year in advance before they submitted[59]. Ultimately, before Pan's conference abstract was released, Boyden and Deisseroth signed a statement claiming to have been the creators of using this channelrhodopsin technique secretly in the lab[59]. Nearly ten years following their filing, in March 2016, the pertinent patent was granted.

Pan was quite late when it came to publication, as it was only after three separate groups had published their papers on channelrhodopsin that he could finally publish his own. He was not a recipient of the two major awards that Deisseroth and Boyden received: the Brain Prize (2013), which awarded one million euros to six optogenetics inventors, and the Breakthrough Prize (2015), which gave Boyden and Deisseroth each $3 million. As a matter of fact, Deisseroth received nearly $18 million in funding from the NIH for optogenetics research since 2005, and Boyden has earned over $10 million[59]. Both of their labs receive additional funding each year thanks to other significant projects. Boyden is a well-known speaker who has delivered several TED talks; in 2015, the New Yorker published a detailed biography of Deisseroth. Currently, Deisseroth co-founded and is also the scientific advisor of Circuit Therapeutics, a business that is utilising his patients to develop a variety of optogenetic medicines. In contrast, Pan has only gotten slightly more than $3 million in total over the previous ten years and currently holds one NIH grant, which is the absolute minimum required to maintain a research programme[59].

On a more positive note, Pan was also awarded a patent for his use of channelrhodopsin to improve vision in the eyes. RetroSense, which received recognition with an award from the Angel Capital Association in 2015, licensed his patent[59]. Pan started clinical trials in 2016 to use gene therapy to deliver algal proteins to blind individuals. It's also the first optogenetics trial conducted on humans and the first gene therapy trial involving non-human genes. Amazingly, in our time, there could be blind individuals in Texas who are walking about with proteins and DNA from algae in their eyes. This was Pan's objective this whole time. Pan remarked, "One thing I still feel glad about is that even right now, our clinical study is still ahead of anyone." However, the path to optogenetics success in humans will probably take quite a while since, in the US, there are currently no gene therapies authorised for clinical use[59]. Optogenetics is a tool that UC Berkeley neuroscience professor Yang Dan uses to study sleep, but even so, she isn't relying on optogenetics treatments to become available in clinics soon. She said, "I believe that these safety checks will take a long, long time."[59]

Regarding the invention itself, some experts say that Pan might not have possessed the grand, visionary idea that Deisseroth and Boyden did. One of the other people credited with the first paper regarding channelrhodopsin in neurons, Stefan Herlitze, stated, "Of course, I have to say, Deisseroth and Boyden, they really developed the field further." Boyden concurred. "Karl and I were very interested in the general question of how to control cell types in the brain," he stated. "In recent years, we worked to push these molecules to their logical limits."[59]

This story suggests that it might not matter who invented optogenetics—instead, what matters is who has pushed the boundaries of science the furthest. Pan refused to respond when asked if he deserved the recognition Boyden and Deisseroth had received. He subsequently stated that Deisseroth "also did a very excellent job, no doubt. But he's also very lucky because if our paper was ahead of him, the story would be different. We would have gotten more credit."[59] The obstacles are not just unique in photonics but also in all scientific advancements, as can be seen through such anecdotes.

Future developments in optogenetics could include achieving single-cell precision and a comprehensive understanding of brain circuitry. Numerous optogenetic investigations, for instance, have engaged neurons simultaneously as an ensemble by activating or silencing them as populations[60]. However, in the brain, even similar-looking neighbouring neurons can exhibit highly different patterns of activity, which brings up the issue of whether optogenetics can be achieved with single-cell-targeting precision[61]. It has been the focus of much optogenetic research to target historically defined cell types, such as norepinephrine-producing or somatostatin-positive neurons, which each contribute to the regulation of cognitive function, arousal and more, as well as inhibitory neurotransmissions, respectively[61]. This has raised the question of whether any desired type of brain cell could be controlled. Of course, there is currently no comprehensive cell taxonomy for nearly every species or brain circuit[61]. It may be feasible to optogenetically activate or silence any particular type of neuron that is thought to be involved in a behaviour or a disease if comprehensive descriptions of cell types are attained, and especially if the selective genetic handles that enable targeting specific cells for gene expression are discovered[61]. This is an illustration of how advancements in biophotonics are making a difference.

Nanophotonics

Actually, modern nanotechnology only truly began in the 20th century. In 1959, American physicist Richard Feynman delivered a lecture titled "There's Plenty of Room at the Bottom," where he introduced the fundamental concepts and principles of nanotechnology. Although Feynman did not explicitly use the term "nanotechnology," he described a method allowing scientists to control and work with individual atoms and molecules. However, arguably, it was the creation of the scanning tunnelling microscope in 1981—which made it possible for engineers and scientists to observe and work with individual atoms— that marked the true beginning of modern nanotechnology. Even now, microscopy remains a significant research area of nanophotonics. One goal is to construct what is known as "superlens", which would create images that surpass the precision of the diffraction limit, the limit of resolution performance[24]. In general, the term near-field microscopy can refer to any technique that uses the near-field, the part of an electromagnetic field closest to the source, to attain subwavelength, nanoscale resolution[24]. This illustrates one purpose for the expansion of photonics technology.

Quantum mechanics

A final topic, a pivotal chapter in the history of contemporary physics, is the development of quantum mechanics. The major historical chapters surrounding this area start with the advancements of quantum theories to account for various researched and observed phenomena during the period, such as blackbody radiation, solar emission spectra, and the photoelectric phenomenon. There were two stages in the formation of quantum mechanics. Around 1900, the old quantum theory's first phase began with radically new techniques to explain physical phenomena that the 1800s classical mechanics could not explain[62]. The story started inconspicuously on the 14th of December, 1900. Speaking to the German Physical Society, Max Planck discussed the continuous range of light frequencies emitted by a theoretically ideal absorber and radiator of heat, referred to as a "black body." About two months prior, the 42-year-old theorist had proposed a formula reflecting some new experimental results. Now that he had more time and freedom to reflect, he sought to offer a scientific explanation for his formula. Planck, idealising it somewhat, compared a bit of matter to an assembly of fluctuating electric charges[62]. Next, he visualised dividing its energy into several portions corresponding to the oscillation frequencies. He decided to refer to the proportionality constant as h. We now write this using the formula e = hf. The oscillation frequencies dictated the emission frequencies of the light. The proposed formula of Planck was subsequently replicated using a convoluted series of logic that now included the same natural constant, h[62].

In retrospect, we might anticipate revolutionary commotion. But things were ambiguous, as they frequently are in history. Planck did not refer to his energy elements as quanta and was not inclined to emphasise their discrete nature, which was not particularly intuitive. This is why only over time did the significance of his process become apparent. That is to say, even though the issue he was treating was significant at the time as we know it now, its effects were initially believed to be limited.

Subsequently, Niels Bohr put forth a revised atom model that included quantised electron orbits in 1913. Although electrons continue to orbit the nucleus like planets orbiting the sun, they can only occupy specific orbits and not orbit at any random distance. Contrary to what would be predicted traditionally, an electron did not continuously travel between orbits around the nucleus when an atom received or emitted energy. Instead, a photon of light would be released as the electron instantly jumped from orbit to orbit. Louis de Broglie proposed a ground-breaking theory in 1924: matter possesses wave qualities. Building on Einstein's idea that the photoelectric effect results from quantised energy transfers and on his other proposal from special relativity that mass at rest is equivalent to energy or interchangeable, De Broglie postulated that matter in motion appears to have a corresponding wave[62]. He expanded on the Bohr model of the atom by demonstrating that an electron orbiting around a nucleus may be considered to have wave-like properties[62].

This made it possible for the three German physicists, Pascual Jordan, Max Born, and Werner Heisenberg, to develop matrix mechanics in 1925, which is part of what gave rise to current quantum mechanics[62]. This is the first mathematical expression of quantum physics that physicists have discovered. Originally, Werner Heisenberg created this formulation of the laws of physics only as an equation to help him predict the photon intensities in the different hydrogen spectrum bands. Similar to this, as a rough representation of the generalised case of de Broglie's hypothesis, Austrian physicist Erwin Schrödinger—you may recognise his thought experiment of Schrödinger's cat—invented the classical non-relativistic Schrödinger equation and wave mechanics[62]. This linear partial differential equation governs the wave function that describes a quantum mechanical system.

Meanwhile, the De Broglie-Bohm theory is a general theory that pertains to the entire cosmos. According to the guiding equation, every particle in the cosmos moves according to a single wave function. Every moving particle can behave as a wave and a particle at different times, according to the De Broglie equation while the matter-wave, also referred to as the De Broglie wave, is the wave that is connected to the moving particles[62]. The de Broglie wavelength is the name given to the wavelength.

Quantum computing

Now that we have covered some of what the term "quantum" encompasses, where does computing come in? The invention of the computer as we know it today was made possible by the contributions of Ada Lovelace (1815–1852) and Charles Babbage (1791–1871)[63]. Although Babbage never got around to creating the Analytical Engine he designed, it has now been established that it would have functioned. In fact, its logical structure is essentially identical to that of contemporary computers. On the other hand, Lovelace is recognised as one of the pioneering computer programmers and the first to understand the full potential of computers[63]. She was the first to realise that pure calculation was not the limit of the machine's uses and published the first algorithm meant to be executed by a computer of that kind[63]. Subsequently, Alan Turing proposed the concept of a universal machine—later dubbed the Turing machine—that could compute anything that could be calculated in 1936. His theories laid the groundwork for the main idea of the contemporary computer[63].

The timeline of quantum computing began in 1982 when Richard Feynman lectured on the possible benefits of quantum computing systems[64]. Different algorithms related to quantum computing developed over the following decades. An algorithm that could determine the factors of huge numbers in a relatively short time was presented by Peter Shor in 1994, now referred to as Shor's algorithm. It outperformed the best classical method by a wide margin and raised theoretical concerns about modern encryption security. Lov Grover then proposed a technique for quantum computers in 1996 that would be more efficient at database searching, and this is now known as Grover's search algorithm. It wasn't until 2010 that D-Wave One, the first quantum computer available for purchase was released (the annealer). Some more notable moments in the early 21st century include 2016, when IBM provided access to quantum computing through IBM Cloud, and 2019 when Google claimed the achievement of quantum supremacy. John Preskill used the phrase "quantum supremacy" 2012[65] to characterise the point at which quantum systems could outperform classical systems at specific tasks.

In 2020, scientists from the University of Science and Technology of China, under the leadership of Chao-Yang Lu and Jian-Wei Pan, used their quantum computer, Jiŭzhāng, to carry out a method known as Gaussian boson sampling[66]. The outcome, published in the journal Science", exhibited 76 photons detected, which is far more than the five detected photons that were the previous record and more than the power of traditional supercomputers[64]. Jiŭzhāng, in contrast to a conventional computer constructed with silicon processors, was a complex tabletop apparatus comprising lasers, mirrors, prisms, and photon detectors[66]. A quantum computer consisting of photons has outpaced even the world's most influential classical supercomputers in a significant milestone. While it is not a general-purpose computer capable of sending emails or storing information, it does show the valuable promise of quantum computing.

As mentioned, the year before, Google made headlines when Sycamore, its quantum computer, achieved the remarkable feat of completing a task that would have supposedly taken a supercomputer an astonishing 10,000 years to accomplish in just three minutes[66]. However, depending on your estimation method, others like IBM argue that the identical task can be perfectly simulated in 2.5 days with much more fidelity on a classical computer system[67]. According to the USTC team's research, it would require an astounding 2.5 billion years for the third-most fastest supercomputer in the world, Sunway TaihuLight, to complete identical calculations as Jiŭzhāng[68]. The phrase "quantum primacy" refers to the scenario in which a quantum computer outperforms a classical one exponentially and accomplishes tasks that would otherwise be almost impossible to calculate, which has only been shown twice[69].

Chapter 3: Technical Information

As mentioned in the introduction, photonics involves the use of photon-based light production, detection, and manipulation[15]. There are a multitude of ways we can control light; we can do so through emission, transmission, amplification, modulation, switching, signal processing, and sensing of light. But how are we able to do so? In the introductory chapter, some components were already explained including lasers and fibre optic cables. However, numerous other components are involved in photonics, from more commonly known devices like lenses to lesser-known ones like waveguides. Before diving deeper into the different specific applications in different fields, it would be useful to know a little about how these technologies can allow us to control light in order to visualise better, so this chapter will go through some more of the main components used in photonics technology. We shall continue with a brief history of each component, but some more technical aspects of the science behind each element will be provided.

Lenses and prisms

Starting with a more basic component, lenses are made of transparent material with curved sides for bending light rays to form images. Lenses can be found in devices like cameras, microscopes, and within your eyes. They are used to focus images, such as how the lens in your eye focuses light rays on your retina to be able to see or to magnify and demagnify objects, as observed in magnifying glasses. Lenses work through the principle of refraction.

There are two types of lenses: convex and concave. Due to their shape, convex lenses converge light rays, meaning they focus light rays to a single focal point to form an image. In contrast, concave lenses diverge light rays, spreading them apart. They are used to form both real and virtual images. Real images refer to those that form when light rays converge at a focal point. Hence becoming possible to project these images onto a screen. Virtual images are formed by the apparent intersections of diverging light rays traced backward. These rays do not converge at a physical location, making it impossible to project the virtual image onto a screen. Projectors produce real images, while devices like magnifying glasses produce virtual images only visible to the user. In glasses, concave lenses can correct nearsightedness, while convex lenses are used for farsightedness. This is because to see closer objects, the lens in your eye needs to converge the light rays more to form a clear image.

Meanwhile, in nearsightedness, the lens in your eyes converges the light too much, so a concave lens is needed to diverge some of the light.

Lenses date back to at least 700 BC when they were made in Ancient Egypt and Mesopotamia. Inspired by the optical effects they observed with water, they began to polish crystals to mimic these abilities. Made in ancient Assyria between 750 and 710 BC, the Nimrud lens is among the most well-known examples of the original lenses. This lens was meant to be a magnifying glass, a tool to start fires or a decorative piece. Then, in 60 A.D., emerald lenses were used by the famous Roman Emperor Nero to watch gladiator games. His jewel lenses quickly became a sought-after accessory for upper-class fashion[70]. Over time, lenses have been used in many optical devices that process light waves, such as telescopes, magnifying glasses, microscopes and cameras. For example, before the 17th century, magnification tools had never been implemented to study extraterrestrial bodies[71]. Therefore, when Galileo used telescopes for space observation and built telescopes with increasingly higher magnifying power, he was able to revolutionise astronomy. Due to Galileo's groundbreaking research, optical telescopes with ever-increasing power have been created, along with other inventions, including the camera, computer and spacecraft, allowing observational capability to be improved even more. Thanks to these advancements, significant progress has been made in scientific understanding of the solar system and the universe[71]. This is one example where it can be seen that while the lens may seem like a simple device, it is a crucial element in the progression of photonics technology.

In the field of optics, a prism is a carefully crafted piece of transparent material, typically glass, with precisely cut angles and flat surfaces[72]. It serves as a valuable tool for examining and redirecting light, and triangular prisms are particularly useful for breaking down white light into its individual colours, known as a spectrum[72]. This phenomenon is called dispersion and is the key use of prisms, as demonstrated by Sir Isaac Newton[73]. Different wavelengths represent different colours; as they each hit the prism, they bend or refract a different amount. Those nearer to infrared have longer wavelengths and are bent the least, while those near the ultraviolet band have shorter wavelengths and are bent the most[72]. Prisms like these are one of the main components of certain spectroscopes, instruments used to analyse light and its emitters or absorbers by identifying the material's structure and identity[72]. The type and function are partly defined by the angle, position, and quantity of surfaces[72]. This application is employed across many other spectrographic and refractometer components[74].

Before that, along with natural prisms like rock crystals at Pompeii, archaeologists discovered glass prisms in the remains of ancient Roman ruins. These findings imply that, from ancient times, people have been able to perceive a range of colours when white light travels through a prism. Rainbows were created in the 13th century using six-sided natural quartz crystals. For a long time, people have questioned the nature of light and colour. For instance, the Greek philosopher Aristotle (384–322 BC) claimed that colours were combinations of light (white) and dark (black). This was widely believed and gave rise to the notion that sunlight was only one type of light, namely white light[75] and that as light passed through a prism, it changed from being pure white to multiple colours. This view persisted even up to the late 1660s before a letter to the Royal Society of London was published in 1672 by Isaac Newton. According to Newton, the coloured rays already existed in the white light and split as they bent at various angles inside the prism[75]. Our current understanding of colour and light originates from the tests conducted by Newton, where he proved that a prism could both separate white light into its individual colours and recombine the spectrum back into white.

Nowadays, the prism is a handy instrument capable of much more than just dispersing light. It can also rotate, orient, reverse, and invert pictures. This includes rotating an image 180 degrees, flipping it over on its horizontal or vertical axis, and even altering its orientation. Prisms are crucial in binoculars, telescopes and other surveying equipment because, through internal reflection, prisms may change the direction of light. For example, they are typically used to erect the image and flip it upright in binoculars or single-lens reflex cameras, as these cameras usually use a mirror and prism system, permitting the photographer to view through the lens and see exactly what will be captured. Furthermore, there is a type of prescription where prisms are integrated into standard lens prescriptions[76]. Eye doctors frequently prescribe prisms to treat a condition called binocular visual dysfunction[76]. This problem makes it difficult for a person to see a single, clear image since their eyes are not fully synchronised and do not work smoothly together. Since prism lenses lack focusing strength, they cannot correct refractive errors. The function of adding prisms to glasses is to bend light before it travels into the eye. Since two images are perceived in double vision because distinct parts of the retina receive light in different ways, the light is guided back to the correct location on the retina in each eye by the prism to be able to generate one single clear image[77].

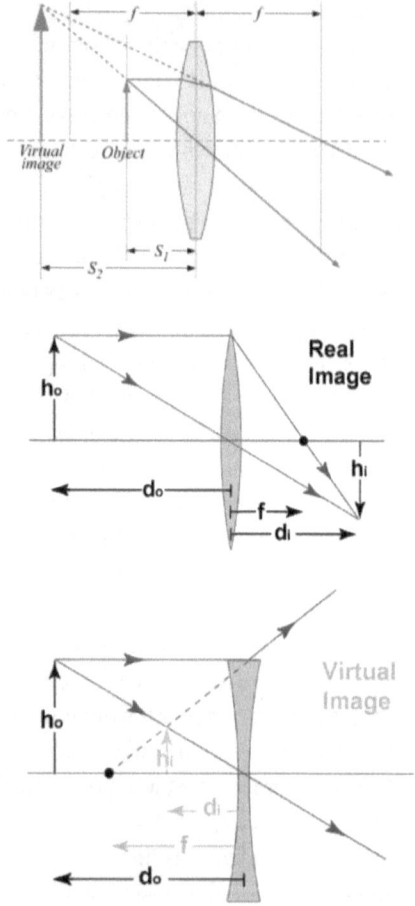

Light ray diagrams illustrating (from top to bottom): a virtual image formed by a convex lens[7], a real image formed by a convex lens[8], light passing through a concave lens[8]

Modulators

An optical modulator is an apparatus for manipulating a feature of light, more often to manipulate precisely a laser beam. Many properties can be manipulated, and the modulators are named after the properties they control, such as phase modulators, intensity modulators, and spatial light modulators. A typical modulator used in photonics is the Mach-Zehnder Modulator, an optical interferometer. As the name suggests, this type of modulator takes advantage of the interference phenomenon of light.

Some main properties of light include intensity, phase and polarisation. Intensity is the power of the optical beam and can be adjusted by changing the amplitude of the light wave, the distance between the peak or trough of the wave and the equilibrium position, or halfway between the peak and the trough. In simple terms, phase means how far or how many cycles the wave has travelled from the starting point. An analogy to this would be a clock, where the reference position for the hands of a clock is at the number 12, and the minute hand has a period of one hour. The minute hand would have a phase of a quarter of the period at 15 minutes past the hour since out of 360° (2π radians), it has traversed a phase angle of 90°($\pi/2$ radians)[78].

Polarisation is another property of electromagnetic waves. Firstly, electromagnetic waves are called such because they are formed from the interaction of electric and magnetic fields travelling through space. These electric and magnetic vibrations occur perpendicularly to each other. There are several planes, or hypothetical flat surfaces, where these electric and magnetic vibrations can occur[79]. Unpolarized light is a type of light wave that vibrates on multiple planes[77]. Light released by the sun or a lamp is an unpolarised light source. To clarify, although waves always propagate or travel in the same direction, the planes in which they occur can vary[79]. Polarisation is the process by which unpolarized light becomes polarised. Polarizers are put over lenses, light sources, or both to reduce hot spots from reflecting surfaces, boost contrast, and remove light dispersion[80]. A common example of this concept being used would be in polarised sunglasses to reduce glare since polarisation affects how a wave interacts with a medium or surface when it interacts with it. For instance, the type of polarisation is essential in satellite system designs to prevent signal deterioration and interference. For example, waves with a linear polarisation reflect differently than waves with a circular polarisation.

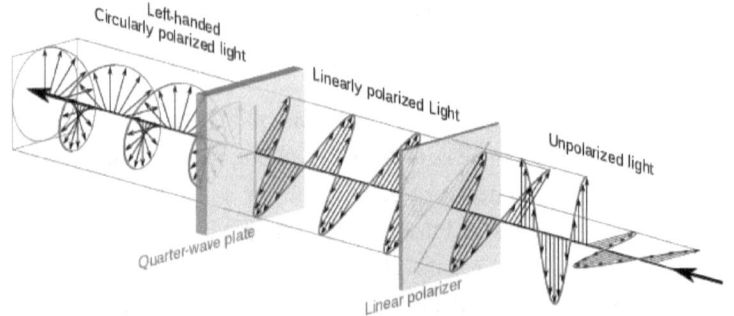

Diagram showing unpolarized light waves and two types of polarisation[9]

Waveguides

Briefly mentioned in the introduction, a waveguide is any device that restricts and controls the transmission or movement of electromagnetic waves. Waveguides come in various shapes and forms, including hollow metallic tubes and optical fibres. A waveguide often has an area with a greater refractive index than the cladding, as discussed earlier in fibre optic cables, although there are other ways of controlling wave propagation. Channel waveguides confine light propagation to a single path while guiding it in two dimensions, and planar waveguides only guide light in one dimension.

The earliest investigations of electromagnetic waves in ducts were conducted by theorists in the 1890s. During this period, Oliver Heaviside vigorously developed the theory of transmission of electromagnetic energy. He introduced numerous concepts that have since been valuable in electrical engineering. It was Heaviside who first proposed the idea that an electromagnetic wave could be guided by metallic conductors in 1893. In 1897, his ideas were advanced further by Lord Rayleigh, but they were not commonly used for any real applications, and Rayleigh's work was forgotten. Afterwards, scientists investigated methods to transmit electromagnetic radiation. Still, it was only almost 40 years later when waveguides were rediscovered by G.C. Southworth and W.L. Barrow, who, for nearly five years, worked independently without knowing each other[81]. Interestingly, it was only until just a few weeks before they were due to present papers on waveguides at a joint meeting of the Institute of Radio Engineers and the American Physical Society in May 1936 that Barrow and Southworth became acquainted with one another's work[81].

Building on even further by bringing in silicon photonics, a leading area of photonics mentioned in the introduction, this area of research started because Richard Soref et al. proposed the idea of using chips with silicon waveguide devices in the mid-1980s, leveraging the well-established silicon microelectronics industry. Because of this proposal, the field of silicon photonics kickstarted, leading to the decades of development and advancements we know today.

Since their purpose is to guide waves as wires would guide electricity, it is clear that waveguides are a crucial element in photonics devices. For example, in the way electronic integrated circuits include numerous tiny wires, photonic integrated circuits also have channel waveguides. The waveguides connect and transport light between different optical components. There is a type of laser that uses waveguides as the gain medium or laser medium. Because waveguides confine light to one direction, the beam divergence is effectively removed, and such high optical intensities or power can be maintained over a long length. Furthermore, in some optical sensors, the channel waveguide is the active sensor element located at the top surface, which is then exposed to external stimuli[82].

Photodetector

Photodetectors are commonly considered light sensors but can also measure other electromagnetic radiation. More specifically, they can be seen as photon detectors, and they work based on the principles of the photoelectric effect. They can convert incident optical and output electrical signals, thus making them optoelectronic devices. Different photodetectors with different sensitivities to various wavelengths are used depending on the specific purpose. For example, solar-blind photodetectors are only sensitive to ultraviolet light but not visible sunlight or infrared.

Photodetectors have a diverse spectrum of applications. They can be useful in measuring optical powers in devices such as spectrometers (devices for detecting and analysing wavelengths of electromagnetic radiation), optical data storage devices and microscopes, laser rangefinders like LiDAR, experiments involving quantum optics like in testing lasers, and gadgets for night vision, extremely sensitive photodetectors are needed[82]. Optical fibre communications require particularly quick photodetectors since the detection of rapidly pulsing lasers is needed. Image sensors are found in imaging applications such as cameras. In more everyday life scenarios, photodiodes, a type of photodetector, are used in remote controls as the receivers of infrared remote controls for televisions, DVD players, and other electronic devices. Smoke detectors also use photodiodes to detect smoke.

Photodetectors are similar to photovoltaic systems as they both use the photoelectric effect. However, photodiodes are optimised for light detection, while solar cells are optimised for energy conversion efficiency. Willoughby Smith identified selenium's photoconductivity in 1873. Its ability to become more electrically conductive when it absorbs electromagnetic radiation. In the early 20th century, researchers began to explore using selenium as a photodetector. Selenium photovoltaic cells were developed by Charles Fritt, as mentioned earlier in the solar cells section, which generated electrical voltage when illuminated, but their efficiency was relatively low. Due to their small scale, high detection efficiency, and quick detection speed, semiconductor photodetectors—also known as photodiodes—are the most popular varieties of photodetectors found in systems for optical communication[83]. Also, as seen in laser diodes, a diode is an electrical component with two terminals that limit current flow in one direction while conducting current mostly in the other[84]. Many photodiodes have been designed over the years, such as vacuum photodiodes, avalanche photodiodes and PN junction photodiodes. Nano photodetectors are a more recent area of development. The increasing advancements in nanotechnology have allowed nano photodetectors to improve as well. These devices utilise nanostructured materials such as nanowires, quantum dots, and 2D materials to enhance light absorption, sensitivity, and response time. Nanophotodetectors offer the opportunity for improved performance and integration with other nanoscale devices.

Photonic integrated circuits

The components mentioned above are some of the components included in a photonic integrated circuit. Skimmed upon in the introduction, Similar to how an electrical switch is turned on to drive electronic components, PICs inject light using a laser source, which powers the optical components. Compared to integrated circuits based on electrical components, PICs process and distribute information via photons instead of electrons, which means photons go through optical components, including phase shifters, polarisers, lasers, and waveguides[85]. Photonic integrated circuits were introduced in the late 20th century and have continued to make rapid improvements, albeit not at the speed of electronic components in the mid-20th century. This is evident as many daily devices use electronic components. These are used in many photonics applications, such as sensors or neural networks.

PICs fall under the category of integrated optics, which is the study and use of optical systems and components integrated into a single, small platform or chip (micron to mm in size), enabling sophisticated functionalities comparable to those of electronic integrated circuits. Typically, photonic integrated circuits are created using wafer-scale technology, which involves building an enormous circuit on a single wafer—a thin disc made of semiconductor material—on substrates, or "chips," made of silica, silicon, or other materials[86]. The substrate is the substance that houses the parts that support a printed circuit board, and the nature of the substrate dictates many of the technology's properties and constraints. Once more, silicon photonics—directly implementing photonic capabilities on silicon chips—is a rapidly developing field. Other materials are accessible, though. For instance, indium phosphide (InP), which is mainly utilised in optical fibre communications, is the foundation of a photonic integrated circuits technology that has already been commercialised[86]. It is challenging to scale and integrate discrete optical components into intricate systems. The deployment of intricate discrete optical designs has hindered the growth of network interconnects to satisfy the demand for data. Discrete optical designs and electronic communication systems have constraints that integrated optics can overcome. Further expansion of PIC functionality for broader market applications will come from future integration with electronic circuits[87].

Going back to the start, the concept of integrating photonic components on a single chip emerged in the 1970s when researchers recognised the potential advantages of combining photonic devices[21], such as lasers, modulators, and detectors, with electronic components on a single platform. The idea for an integrated optical circuit was initially presented by Stewart Miller of AT&T Bell Laboratories[88], the research division of an American multinational telecommunications holding company in 1969. This integration promised to enhance functionalities, improve efficiency, and enable new applications. Between the late 1980s and the early 1990s, integrated optics became a mature technology, and optical communications were increasingly the motor of its development. Researchers began to demonstrate the integration of individual photonic components on a common substrate[21]. Big industrial groups made strong efforts in silicon-based, lithium niobate, and even glass technologies. Then, the late 1990s, marked by the dawn of the internet revolution, can be seen as the golden era for optical communication and integrated optics technologies.

In the late 2000s and early 2010s, the commercialisation of PICs gained momentum. Monolithic InP-based PICs, first introduced in 2004, established commercial viability for the mass production of integrated photonics for telecom networks[21]. Companies started offering silicon photonics-based products and solutions for multiple industries, such as data communication, telecommunications, sensing, and biophotonics. PICs found applications in data centres, high-speed optical networks, optical interconnects, and other areas requiring high-performance and scalable photonic technologies, enabling simpler, more reliable, cost-effective, higher bandwidth communications.

As PIC technology progressed, more investigation was done into combining photonic and electrical components. This integration involved combining photonic and electronic components on a single chip[21] and developing methods to enable efficient communication and data transfer between the two domains. As InP PICs grow more intricate, wiring density between photonic and electronic parts and heat generation becomes increasingly important considerations. Hybrid integration, co-packaging, and co-design approaches were explored to enhance performance and functionality. Co-packaging involves placing multiple components in a single package, emphasising the physical integration of components, often with a shared substrate. Co-design takes a holistic view involving the simultaneous design and optimisation of multiple components. This approach considers the interactions and interdependencies between these components to create a more efficient, reliable, and high-performance system.

To date, the most intricate and sophisticated PICs contain over 1000 optical components on single, monolithic, InP-based chips. The application of membrane-based photonic technologies creates a roadmap for integrating over 10,000 components per chip. It provides the necessary space and energy savings for closer integration with electronics and higher-density integration[87].

How do different PIC materials compare? As mentioned in the introductory chapter, the main drawback of silicon photonics is that they lack the intrinsic ability to produce laser light. In contrast, InP can both produce and "conduct" light, making it the only base material out of the three common materials (Indium Phosphide, Silicon Nitride and Silicon Photonics) that is inherently able to integrate active components like detectors and amplifiers. This explains why InP dates back the longest in commercial history[89]. Silicon Nitride(SiN) and Silicon Photonics-based systems require an external laser, sometimes using InP. However, silicon nitride has an edge over InP-based PIC in terms of possessing a spectral range from near-infrared to UV, making it suited for applications like biomedical sensing or augmented reality where visible light is used[89]. InP is unable to process visible light. The low light losses of SiN are another advantageous feature as more complex circuitry can be created[89]. Although silicon photonics cannot overtake these other materials in these characteristics, its main advantage of being able to be produced at already existing electronics circuitry fabs (semiconductor production facilities) means it still has important applications[89].

The full potential of integrated photonics can only be unlocked by incorporating each platform with each other. The term hybrid integration mentioned earlier simply means manufacturing the required individual components separately and then integrating them into higher-level assemblies[89]. Without precise waveguide alignment and high-quality connections, light loss would be a potential obstacle to hybrid integration. Conversely, it is highly flexible as manufacturers can choose the appropriate foundries for each component to be assembled later. Users can also simply buy different components off the shelf to integrate together. However, in more complex or high-volume applications, another possible method is heterogeneous integration[89]. This is where the materials are actually combined. For instance, a laser can directly be built on silicon by fusing a small piece of InP with silicon. This method is less developed, and the developing stage may be more expensive, but it will pay off and be more cost-effective for high-volume applications[89]. For now, hybrid integration remains a staple in the integrated photonics industry.

Optical switches, amplifiers, filters, couplers and splitters

Optical filters are for transmitting light selectively within a specific wavelength range or colour spectrum while absorbing the remaining light. Optical filters are widely employed in various optical instruments to change the stage lighting's colour and in photography, wherein occasionally both absorptive and special effect filters are employed to analyse infrared radiation without causing film or sensors to be interfered with by visible light and overwhelming the desired infrared, astronomers use optical filters to limit light passing to the spectral band of concern. In fluorescence applications like fluorescence spectroscopy and microscopy, additional optical filters are necessary.

This may sound relatively simple, but there are still different types of these filters. The most straightforward type would be the absorptive filter, which simply absorbs all unwanted wavelengths while letting desired ones pass. More physically complex types would include interference filters. One example would be thin-film filters. A thin-film filter comprises several thin dielectric material layers with various refractive indices. They are called dichroic filters and work on the thin-film interference principle, similar to how oil films on water's surface produce colours[90]. When an oil film is illuminated at an angle, a portion of the light is reflected off the top surface and a portion of the bottom surface when it touches the water[90]. The colours observed result from some light wavelengths being strengthened due to this delay, while other light wavelengths tend to be cancelled since the path taken by the light reflected from the bottom is a little bit longer.

A dichroic filter builds alternate layers of optical coating on a glass substrate rather than an oil film to create interference. They have different refractive indices on a glass substrate as opposed to the oil film idea. Phased reflections are produced at the interfaces between layers with different refractive indices, which selectively reinforce some light wavelengths while interfering with others. The filter's frequency passband, the spectrum of wavelengths or frequencies that can travel through it, can be adjusted to be as broad or narrow as needed by adjusting the number and thickness of layers. Energy is not wasted by dichroic filters during operation because undesirable wavelengths are not absorbed but rather reflected. As a result, compared to a typical filter that aims to absorb all radiation except that in the passband, not nearly as much heat builds up in them.

An optical switch performs the same purpose as its electrical equivalent. An electrical switch is a component that controls the flow of electric current, either allowing or preventing its flow[91]. Electricity typically tries to go from one place to another. A continuous wire is needed for electricity to continue to flow. In other words, electricity cannot flow if a wire is broken. A switch is the apparatus that enables the joining of wires. The crosspoint contact conducts electricity. For example, when you press the key in a mechanical switch, two metal parts of the switch make contact with one another, thus registering a key press[91].

On the other hand, an optical sensor detects the induction of a light signal that triggers optical switches to register a key[92]. Certain systems function by first transforming incoming light signals into electrical signals, analysing them, and returning them to light signals. In summary, an optical switch is a device capable of selectively switching light signals flowing through integrated optical circuits or optical fibres connected from one circuit to another. Put differently, light signals can be transferred across various communication network channels using optical switches[93].

Several methods are used by optical switches to manipulate these signals. One such technique is to guide the light signals using tiny, mechanically controllable mirrors or prisms. This type of device, where the mirror position may be electronically adjusted to direct the light path, is called a micro-electro-mechanical system. Additionally, there are bubble switches that work by producing and managing a small gas bubble inside a liquid-filled chamber. The bubble directs the signal by controlling the refraction of light. Other techniques involve controlling the light path by modifying the optical medium's physical characteristics. Among these are thermo-optic switches, which heat a waveguide to change its refractive index and the direction of light flow. Another example would be electro-optic switches, which leverage the electro-optic effect, in which the course of the light signal is restricted by applying an electric field to alter the optical medium's refractive index.

In the fields of electronics and telecommunications, coupling refers to the process of transferring electrical energy between two or more circuits or within different components of the same circuit[94]. Optical coupling serves the same purpose; this technology uses light waves to transmit electrical signals, maintaining electrical isolation between separate circuits or systems while coupled[95]. Electronic isolation prevents the undesirable exchange of direct and alternating currents between two circuit sections[96]. Optical fibres must be aligned with another fibre or an optoelectronic device like LEDs, modulators, or laser diodes[97]. This alignment process can be achieved by precisely positioning the fibre in direct contact with the device or using a lens to facilitate connection over an air gap.

As the name suggests, optical splitters take a single light beam as input and divide it into multiple beams or, conversely, merge multiple beams into one, featuring multiple input and output ports[98]. The optical splitter has been a crucial component in passive optical networks, enabling sharing a single PON interface among numerous subscribers[98]. PON uses fibre-optic technology to deliver data from a single source to multiple endpoints. The term "passive" describes a network where optical fibre cables are linked to a splitter that operates without needing an external power source[99].

Additionally, an optical amplifier boosts the light or optical signal without needing to convert it to an electrical signal, which is vital in supporting modern long-distance optical communication networks[100]. Lasers utilise optical amplification to produce a beam, as mentioned earlier, and the amplification process in fibre optic amplifiers relies on the same phenomenon that is "stimulated emission." When an incident photon stimulates an atom in a doped optic fibre, a fibre that has had an intentional introduction of impurities to modulate its electrical, optical, and structural properties, it triggers the emission of an identical photon. This process multiplies the number of photons, thereby amplifying the signal.

Chapter 4: Biophotonics (medicine)

Now, let's take a more focussed lens on photonics and delve into more details on specific areas within photonics developments. One fascinating and unmistakably dominant field is biophotonics, which involves studying natural and bioengineered material through an optical lens[101]. Being an interdisciplinary study, it combines elements of biology, chemistry, physics, and engineering to bring forth innovations that meet today's standards. This requires the collaboration of people from many different professions, including biologists, biomedical engineers, computer scientists, chemists, medical physicists, pharmacologists, and clinicians of all specialities. From sensing and imaging to detection to treatment, photonics has a valuable role and holds promise for breakthroughs in fields like medicine much further than we could previously envision.

Among the most prominent areas of photonics are sensing and imaging technology. In general, biosensors are ubiquitous instruments in biomedical diagnostics and play a crucial role in multiple scenarios, including drug development, environmental monitoring, food control, biomedical research, and point-of-care monitoring and treatment[102]. Biosensors can be created using various methods and come in many shapes and forms. Sensitive and precise detection for several analytes can be achieved by coupling these sensors with high-affinity biomolecules[102]. Affinity measures the degree of attraction between a receptor and its ligand. Cellular receptors are proteins found within or on the surface of cells that receive signals. A ligand is a chemical message a cell releases to communicate with other cells. These protein ligands function through binding to a protein receptor.

To combine these concepts, high-affinity binding is the term that describes the longer residence time at the binding site brought about by stronger intermolecular interactions between a receptor and its ligand. The three elements that make up a generic biosensor are the transducing element, the recognition element, and the target. Going one by one, the target refers to the analyte molecule that must be detected or identified by a recognition element employing particular interactions to capture it[102]. Undertakes a physical or chemical change. This change is then converted using a transducer into a signal that can be read.

These devices are particularly useful in bioengineering, and multiple sensing methods are available. A diagnostic technique called molecular imaging uses in vivo observations of cells and molecular structures, meaning inside the living body of a plant or animal, as opposed to in vitro, which would be outside the living body and in an artificial environment. This can be accomplished using different methods of image generation and novel markers for therapy and diagnosis. Therapeutic or biomarkers are naturally occurring molecules, genes, or characteristics which allow for the identification of particular pathological or physiological processes or diseases. They can be found in tissues, blood, or other body fluids and show signs of both normal or abnormal processes or a disease or condition. Molecular, histologic, radiographic, or physiologic characteristics are all types of biomarkers. Put more simply, anything from your weight, blood pressure, and pulse to x-ray findings and genetic tests of blood are types of biomarkers. These biomarkers can help evaluate physical responses to a treatment for particular conditions. Molecular imaging allows disease progression and molecular and cellular pathways to be depicted with reference to a living object. This means we can study biological processes in their physical authentic environments, transcending the restrictions of ex vivo or in vitro biopsy and cell culture laboratory techniques. With these advantages, several clinical goals spanning the entire treatment process can be facilitated, such as optimising treatment plans for specific molecular targets, therapy prognosis, early detection, and monitoring of reoccurrence. Biotechnology companies also implement these technologies to boost the process of discovering and validating drugs.

Evidently, molecular imaging has been beneficial to the medical field, but how does photonics come about? Spectroscopy, a field mentioned earlier, studies the methods of measuring and interpreting electromagnetic spectra. It studies the absorption and emission of light or other forms of radiation by matter[103]. It involves light division, or any electromagnetic radiation to be more precise, into its component wavelengths or a spectrum, like the way a prism disperses light into a rainbow spectrum[104]. Scientists can examine light and study the properties of interacting materials using a spectrometer to record these wavelengths of colour. By measuring the light intensity, wavelength, or colour, scientists can work backwards to deduce the characteristics of the materials the light touched along its route, including object composition, velocity, size, and temperature[105]. Like everyone has unique fingerprints, each type of molecule and atom will reflect, absorb, or emit electromagnetic radiation in its own characteristic way. Therefore, spectroscopy can be effective in chemistry in detecting, differentiating, or quantifying a sample's structural and molecular properties. This principle can be applied to biological contexts for molecular imaging in living organisms.

Magnetic resonance spectroscopy/imaging (MRS/MRI)

One example of where the application of spectroscopy can be highlighted is magnetic resonance (MR) spectroscopy. This diagnostic and analytical tool allows for the non-invasive measurement of biochemical changes occurring in the brain[106]. It is mainly used to determine tumour type and aggressiveness and distinguish between tumour recurrence and radiation necrosis, which is the permanent cell damage as a consequence of radiation therapy. MR spectroscopy can be useful in obtaining in situ concentration measures for certain chemicals in complex samples, such as the living brain. This is done on the same machine as another conventional technology, magnetic resonance imaging (MRI). Composing a computer, a powerful magnet, and radio waves, an MRI scan is used to create highly detailed and insightful images[107]. The distinction between the two is that while MRI can identify the anatomical position of a tumour, MR spectroscopy allows the information to contrast the chemical makeup of aberrant tumour tissue and normal brain tissue[108]. MR spectroscopy can also be beneficial for stroke and epilepsy patients to detect changes in tissues[109].
It should be noted that both MRIs and MRSs utilise radio waves. Although the central focus of photonics is the electromagnetic spectrum's visible region, more importantly, visible light, it can concern all parts of the spectrum, with radiofrequency photonics being another area of research. In short, spectroscopy is a separate and different set of tests in addition to the MRI of the brain or spine used to assess a suspected tumour's chemical metabolism.

MR spectroscopy can analyse hydrogen ions or protons[107], just to name some molecules, by detecting radio frequency electromagnetic waves generated by a molecule's atomic nuclei. Hydrogen nuclei are analysed due to them being the most prevalent nucleus in tissue, and since it's the nucleus that is analysed, this is also known as nuclear magnetic resonance (NMR). The term "metabolism" describes all of the ongoing chemical reactions inside your body to support life and regular functioning, specifically, to provide energy for your body. During this process, several types of metabolites can form, such as amino acids, lipids, and lactate, which can aid in differentiating between tumour types based on their measurements. NMR spectroscopy has been a cornerstone in physics and chemistry for at least 50 years, and MR spectroscopy is essentially equivalent. MR spectroscopy is when NMR spectroscopy is applied in a medical or biological context. Such terminology that eliminated the inclusion of the word nuclear was explicitly adopted to avoid misunderstanding that nuclear radioactivity is involved in these procedures.

Here's a short history of what has become a standard technology in medical practices today. In 1919, Isidor Isaac Rabi graduated from Cornell University with a degree in chemistry, but in reality, chemistry was not his true passion. He spent three years not accomplishing anything until he decided to go to graduate school at Cornell in physics instead[110]. Rabi then organised his molecular beam lab and began looking into the challenge of figuring out sodium's nuclear spin and related magnetic moment, which is, in other words, its magnetic strength and orientation. Across the 1930s, Rabi updated increasingly accurate results in his studies by improving the molecular beam method. In 1937, Rabi hypothesised that it was possible for magnetic moments of nuclei in these tests to be induced into flipping their orientation given enough absorbed energy from an electromagnetic wave with a specific exact frequency. This energy level would subsequently be emitted, causing it to return to the orientation with lower energy. It was this transition between states that Rabi could detect. In 1938, by passing a molecular beam across a magnetic field, Rabi discovered that it was possible to cause them to release radio waves at certain frequencies, and thus, he discovered NMR. He named this technique molecular beam magnetic resonance, and in 1944, he was given the Nobel Prize "for his resonance method for recording the magnetic properties of atomic nuclei."[110]

Nearly a decade later, Felix Bloch and Edward Purcell each worked independently to find a method to study atomic and molecular magnetic resonance properties in both solids and liquids rather than singular atoms or molecules, which Rabi had experimented with[110]. Their work subsequently enabled MRI scans to create images using the body's water content. Purcell and Bloch were also awarded the 1952 Nobel Prize in Physics. Furthermore, this technology was being researched worldwide, as seen by two Swedish researchers, Gunnar Lindström and Erik Odebladalso, who investigated NMR's uses and showcased their research paper in 1955[111]. Their theory was that due to the distinct ways tissues absorb and organise water molecules differently, there were variations in the responses of biological tissues and water[111].

However, it wasn't until 1969 when physician, professor, and scientist Dr Raymond Damadian proposed that magnetic resonance had the potential to be specifically specialised to discern between cancerous and non-cancerous cells[111]. He was able to successfully demonstrate his hypothesis using rats. Damadian discovered that they had differing response lengths depending on the kinds of tissue emission, with cancer cells having longer response signals. It was also Damadian who came up with the notion of combining MRI and NMR. His study into potassium and sodium content in living cells propelled him to his initial experiments with NMR[111]. With his realisation of the capabilities of NMR to differentiate healthy and cancer cells and tissue due to the varying time of relaxation in comparison, an MRI body scanner was first proposed in 1969[111]. Damadian submitted the first MRI patent application in 1972. Damadian designed and constructed a full-body MRI machine right after his patent was approved in 1974[111]; the first MR scanning machine named "Indomitable" now resides in the Smithsonian, the world's largest museum and research complex in the USA. Incredibly, the first-ever human NMR image was achieved on July 3rd 1977. It was a cross-section of the chest of his postgraduate assistant. The image provided visuals of his lungs, vertebrae, heart, and musculature[111]. This was what started the technique called MRI. This breakthrough meant that the use of MRI was no longer restricted to the field of research, with applications and clinical practice only beginning to become increasingly common. The FONAR Corporation was founded in 1978 by Damadian. In 1980, FONAR was the very first company to produce commercially accessible MRI machines[111]. Now, they are frequently employed for imaging inside body structures, exceptionally soft tissues such as the brain. Nowadays, in the US, the company owns 15 facilities for MRI scanning.

Looking back, in 1985, FONAR brought out the original mobile MRI often deployed in the ICU where patient movement poses a risk, inside an ambulance or in an emergency disaster[111]. MRI continued to be a prominent technology, as seen in 2007 when the Invention of the Year was given to FONAR by the Intellectual Properties Owners Association Education Foundation. Also, working alongside Wilson Greatbatch, Damadian developed an MRI-compatible pacemaker[111]. Coming full circle, Rabi was able to experience being imaged by an MRI machine before his death in January 1988. "It was eerie. I saw myself in that machine," he stated. "I never thought my work would come to this."[112]

Especially with some recent advances in MR technology, MR spectroscopy is experiencing a resurgence. One particular use of MR spectroscopy is in neuroscience, which may be used to monitor brain activity. It is instrumental in measuring task-related and pathology-relevant dynamic regional variations in neurotransmitters[113], chemicals such as glutamate that allow neurons to communicate with each other. One fundamental property that can be improved in machines of this kind is temporal resolution. While spatial resolution is how well a machine can identify two areas of the brain, temporal resolution focuses on differentiating two events concerning time[114]. It answers the question of how long it takes to return and obtain data for the same exact location[115]. To cope with the rapidly varying continuous patterns of brain activity that occur with any mental function, a high temporal resolution is of utmost importance[116]. For example, this allows the monitoring of glutamate modulations within the timeframe of less than a minute during cognitive motor and perceptual tasks[117]. Compared to blood oxygen level differences-based MRI, this particular method of in vivo MR spectroscopy, also referred to as functional MRS (fMRS), provides a more immediate measure of relevant neuroactivity and is significantly less prone to being impacted by vascular variations in the circulatory system[113]. This event-related approach allows for the dynamic (real-time) read-out of metabolite changes that are thought to reflect functional changes in neural activity or as a response to changes in the environment by taking data at different points in time. This enables functional MR spectroscopy to carry out non-invasive studies of task-related glutamate variations that provide beneficial information for the study of standard as well as impaired cognitive performance in addition to psychiatric disorders[113].

Early fMRS studies conducted with the 1990s MRS systems evidenced a decrease in glucose and an increase in lactate within the visual cortex[117], the part of our brain responsible for our vision, during visual stimulation. Over time, visual stimulation, along with many other different regions of the brain, has been studied. Also, the latest emergence of high-field MRS systems has drastically revitalised the MRS field. This means we have developed higher strengths of the MR magnet, called "field strength." Circling back to neurotransmitters, glutamate changes can be seen when conducting motor tasks, thermal regulation, experiencing pain, learning, memorising and more. By developing a method to measure these changes, experts can better evaluate the impacts of pharmacological therapies, cognitive intervention, or manipulating physiological conditions, to list some examples. This is important in observing the connection between glutamatergic system activity and memory performance. This is because glutamatergic dysfunction has been hypothesised to be the fundamental occurrence in cognitive ageing, including the neurodegenerative disorders that come with it, such as Alzheimer's disease, and it is also seen in severe psychiatric conditions, including schizophrenia. Therefore, observing any impairment in task-related glutamatergic modulation could be an early marker to detect potential cognitive dysfunction.

Moreover, it can be employed as a monitoring device for treatment responses focussed on mitigating the specified cognitive decline[113]. Observing real-time variation in glutamate levels while carrying out perceptual, cognitive and motor activities may allow new insight into neural bases of abnormal and normal cognitive cognition and behaviours[117]. To achieve this, the apparent changes in this crucial neurotransmitter taking place in the brain must be related to molecular and cellular processes taking place in the brain. With further advancement, fMRS can provide an extremely effective and precise instrument for researching the neurological underpinnings of cognitive processes directly linked to the particular deficits found in psychiatric, neurodegenerative diseases or disorders associated with increased age[113]. It has been proposed that this technology be taken a step further by using optical resonance imaging. This would be a direct analogy to MRI except using optics, though not as much research has been done as of now.

Biophotons

Many living creatures, both animals and plants, such as fireflies and jellyfish, are known for their ability to self-produce light, typically with bacteria[118]. However, interestingly, despite being extremely hard to detect due to its weak intensity, practically every living creature emits a certain degree of light in the ultraviolet or low visible light range[118]. This is because, in addition to releasing energy and generating heat, chemical activities in your body also release tiny amounts of photons, referred to as biophotons. The biological glow we possess is 1000 times dimmer than what a human eye can detect, so the only way we can observe it requires sophisticated devices[119], and scientists can use these observations to their advantage to study different parts of our body.

This discovery dates back to the 20th century. Traces of biophoton therapy research can be attributed to the Soviet scientist Alexander Gurwitsch. His first observation related to the biophoton was in 1923; it was an onion root emitting ultra-week light. He referred to this phenomenon as "myogenic radiation" since he thought these emissions were the product of morphogenetic fields dictating logical development. Morphogenetic fields are the postulated idea that anatomical control is ultimately determined by a hypothesised biological field that retains patterning information throughout cancer suppression, regeneration, and embryogenesis. These fields supposedly contain the required information to influence a living thing's precise form and shape its ability to interact with other beings. When he published his findings, there was quite a commotion, leading to multiple European lecture tours. Unfortunately, when the findings were attempted to be reproduced, they didn't work. Between 1920 and 1935, hundreds of papers on the problem of mitogenic radiation were published, but none proposed a concrete, unanimous conclusion[120].

Whether this type of radiation existed couldn't even be agreed upon[120]. In the Soviet Union, these studies are still carried on but essentially ended in the US and Britain in the 1930s after much futile effort since it seemed impossible to recreate the weak UV light or detect radiation from rapidly dividing cells with photoelectric or biological detectors[121].

And so, with the lack of concrete evidence, the stir that Gurwitsch's work initiated faded until the early 1970s, that is, when German biophysicist Albert Popp rejuvenated it. Popp was researching cancer when he noticed peculiar optics characteristics of a substance found in cigarette smoke and coal tar. He questioned whether these particular characteristics could be the reason for the carcinogenicity of the substance and its ability to cause cancer. It wasn't long until he looked into Gurwitsch's work with mitogenic radiation and discovered the potential of triggering substantial changes in cell behaviour through ultra-week light emissions. Popp coined the term "biophotons" to describe these emissions. Popp stated that "man, essentially, is a being of light."[122] This sums up the idea of biophotons and how we can use them to our advantage.

Popp's ideas were heavily debated within the epidemic circles, but some fully embraced them. Johan Boswinkel was translating Popp's work in 1982. He was fascinated by the concept that our physical, mental, and emotional health is governed by the faintly emitted light from our bodies. In fact, he came up with an incredible leap of a hypothesis:

"If all the information required to control the body's biochemical processes is in the light that the body emits, and if disturbances in that light disrupt biochemical processes and cause disease — as Popp claimed — then it must be possible to 'examine' the light and remove the disease. Then, you return the 'repaired' light to the body. If it works, it will have enormous consequences for everything."[123]

From this proposal, he made a fibre optic device that conducts biophotons to and from the body while continuously improving the light quality with the biofeedback system[124]. Additionally, he founded the Institute for Applied Biophoton Sciences to conduct research and offer training in biophoton therapy. With all this being said, this type of technique is seen more as a pseudoscience, not accepted by all scientists. However, something valuable we can see is its evolution from the 1920s, when it was an unreproducible experiment, to today, where it has become a worldwide industry. Its growth is due to people who claim to have seen benefits from biophoton therapy, which is now a part of holistic medicine and is still not accepted by everyone as a valid treatment method. In conclusion, although it is debatable whether biophoton therapy is entirely scientific, this story demonstrates growing interest in biophotonics in the 20th century.

Closer, in 2009, Masaki Kobayashi finally photographed the dim human glow with the help of an incredibly sensitive camera, which could detect even the faintest bit of light[125]. The glow he observed has a regular pattern, peaking at 4 pm and then brightening and dimming throughout the day. Kobayashi believes that an internal body clock causes this repeating rhythm since he was able to interrupt the pattern and prevent the volunteers' bodies from glowing by forcing them to stay awake in bright light, thereby altering their sleep schedule post-op[125]. It is important to note that the flow of biophotons is not simply a reflection of body heat. Using an infrared camera, it is shown that some of the warmest body areas emit very few photons and that there was no discernible correlation between the total light emission and variations in body temperature[125].

However, our observations collection mostly come from single case studies done so far. Some examples of the information we have include the influence of age and gender on emission, the intensity of emission and its left-right symmetry in illness and health, and emission in different consciousness studies. As mentioned earlier, an important point to remember is that there is yet to be substantial evidence to support biophoton therapy. However, that is not to say there is no evidence of its potential. For example, some research has suggested that light therapy could benefit anaemic patients by stimulating red blood cell production.

Published in the Photomedicine and Laser Surgery Journal, a study discovered an increase in red blood cell production in anaemic rats under red-light therapy[126]. This effect was attributed to the stimulation of erythropoietin, a hormone promoting red blood cell production. These findings were corroborated by a separate study from the Lasers and Medical Science Journal, which found that chronic kidney disease patients experienced improved symptoms of anaemia after undergoing low-level laser therapy[126]. The researchers proposed that the treatment may have enhanced red blood cell production. Some researchers argue that the case studies done so far, others limited, provide enough support to justify further research. Nevertheless, as observed, there is still a lack of concrete evidence to support this technique, so while these studies may suggest that light therapy is a potential candidate for anaemia treatment, more research must be done if these findings are to be confirmed and if they are, determination of the best parameters for effective therapy is needed. Perhaps in the future, anaemia won't need to be treated with conventional methods like iron supplementation and blood transfusions; instead, it will be treated by biophotonic therapy.

Fluorescence

Fluorescence, as mentioned earlier in the discussion of optogenetics, is a concept similar to biophotons, but they are not the same. Fluorescence describes the phenomenon of electromagnetic radiation absorption, most often from the visible light to the ultraviolet range, by a molecule followed by an emission of a lower-energy photon[127]. In simpler words, with molecules, fluorescence would be the glowing of the molecules when hit with specific wavelengths of radiation. This is different to the idea of biophotons in that biophotons are simply natural emissions of photons from metabolic processes, while fluorescence takes in energy from incident radiation. Any small molecules, including lipids, proteins, and nucleic acids, can be labelled with an extrinsic fluorescent molecule, also known as a fluorophore, which can come in the form of another small molecule like a protein or a quantum dot acting as dyes that respond distinctly to light[128]. Due to the distinct characteristics of each fluorophore, the determination of the type of fluorophore optimal for a particular application or experimental system will be affected. In other words, researchers must decide which colour variants and types to use when choosing a fluorescent molecule for imaging.

Conversely, some cells have naturally fluorescent small molecules or proteins, known as intrinsic fluorescence or autofluorescence[128]. When biological substrates like chlorophyll in plants are excited by light at a suitable wavelength, they emit light from the near-infrared to near UV-visible spectral range. An example of such a protein would be the green fluorescent protein (GFP), produced by the jellyfish Aequorea victoria, isolated in 1960[129]. However, its application in cell biology started in the late 1990s, with 1992 being the first instance of cloning. Cloning involves producing replicas of a certain gene of interest since DNA and RNA control the protein production and expression in organisms. GFP is the most common fluorescent protein used, and it is named after the green fluorescence it exhibits after light in the blue to UV range reaches it[130]. They are also compact and chemically inert, making them a suitable indicator for gene activity, labelling proteins and subcellular compartments in live cells. GFP-labelled cells can be monitored in tissues and adapted to identify and develop fluorescent markers in other colours with variants such as yellow fluorescent protein and cyan fluorescent protein. This range of colours is beneficial for applications like multi-labeling of samples.

Sea anemones naturally contain a red fluorescent protein similar to GFP[128]. The first genes of such proteins, including dsRed, were actually isolated and named after the Discosoma sea anemones[128]. Again, with the help of protein engineering, we have extensively widened the range of colours of GFP-like proteins, especially near the far-red end of the spectrum[128]. For example, mCherry, derived from DsRed, is one of the monomeric red fluorescent proteins (RFP) from the mFruits family[131]. For biological material images, these red fluorescent proteins provide the advantage of greater penetration of long wavelengths into cells and tissues. This is because the wavelength of light decreases as you go across the spectrum from infrared to UV, and shorter wavelengths scatter more, leading to less penetration capabilities. RFPs also decrease cellular autofluorescence capabilities in the red range. Usually, it can be challenging to separate wanted from unwanted fluorescence when trying to analyse data on the spectra based on the dyes, probes and proteins we are interested in. That is why decreasing autofluorescence can be advantageous; therefore, red fluorescent proteins are proven to be helpful for whole-body imaging[128]. An instance of this technology being applied would be monitoring the progression of tumours in mouse models[128].

The fluorescent proteins are an example of biological fluorophores. Another form of fluorescence available is the organic dyes, which biologists have used since Adolf Baeyer announced the discovery of fluorescein in 1871. For instance, the most typical fluorescent dye for counting bacterial cells is acridine orange[132]. It interacts with the protein and DNA constituents of the cells to stain both the living and dead[132]. When excited near UV light, the stained cells create an orange fluorescence[132]. In certain situations where metabolically diverse populations coexist, providing the necessary conditions for determining the number of microbial populations by viable count procedures is impossible. For instance, the spread plate or pour plate method depends on a bacterial colony growing on a nutrient medium until it becomes visible enough to count the colonies with the naked eye[133]. In contrast, by using the fluorescent stain instead of observing by eye, underestimation can be avoided, so it is beneficial for counting the overall number of microorganisms in samples, such as soil and water.

Quantum dots are another type of fluorophore. These are semiconductor nanoparticles that exhibit size- and composition-dependent optoelectronic capabilities. Their diameters range from 2 to 10 nanometers (10-50 atoms). Much like the other types that were previously stated, these can absorb light energy, become excited, and emit light photons when they return to their ground state. The Q-dot's size affects the wavelength of light it produces; the smaller the Q-dot, the shorter the emission wavelength.

The fluorescent proteins provide abundant in vivo non-invasive uses, making them extremely useful for protein labelling, allowing precise cell tracking experiments and in vivo whole organism studies on gene expression, protein expression or molecular interactions within a living cell. It is commonly used as a fluorescent reporter, acting as a labelled tag for genes, cells, or organelles of interest, and is usually detected via fluorescence spectroscopy or fluorescence microscopy. That is, the wavelength of the emitted radiation of the molecule is detected to identify the molecule or processes occurring related to it. Relating to this use specifically, mCherry carries the advantage of speed; its quick maturing speed allows for rapid results after transfection[134], the process of deliberately introducing naked or purified nucleic acids (DNA/RNA). It can light up tumours, cancer cells, nerves, and other anatomical features, coupled with microscopy, imaging probes, and spectroscopy for easier and non-invasive identification and precise removal or preservation during surgeries. Fluorescence also offers the advantage of real-time imaging as it eliminates the need for time-consuming image reconstruction and post-processing and enables simultaneous imaging of different molecules or structures, known as multiplexed imaging[135].

There are various ways of labelling structures with fluorescent proteins. One example would be creating unique viruses that secrete the GFP genetic code into the selected cells, typically with coding a specific protein of interest if cell modification is the intended outcome[136]. If present, The target protein is produced by the infected cells along with GFP[136]. The GFP can be produced independently of the target protein or as a single, lengthy molecule with it. The glowing GFP then functions like a tiny beacon that a microscope can follow[136]. Scientists can even pinpoint the precise position of the desired protein when it is fused with GFP inside the cell[136]. Furthermore, GFP fluorescence can be used to monitor cell movement and their changes over time and to determine if a protein was effectively inserted into a cell.

Alas, other than detection, the present use of fluorescence imaging in clinical settings is restricted to a few uses aside from detection, like image-guided surgery for tumours, which was first reported to be performed in 1948 for intracranial tumour neurosurgery, and retinal angiography procedures, which involve blood vessel imaging. This is due to the shallow tissue penetration from the absorbed and scattered. Still, it can be utilised in a preclinical setting to look into disease causes and test new treatments and diagnostics before they are used in clinical settings[137]. This is especially true with the increasing development of fluorescent probes.

In preclinical stages, it can be implemented to study effects on single-celled and multicellular organisms. For example, it can monitor protein expression in vivo in models such as zebrafish. mCherry can label an area or organ in red, creating a single transgenic fish or double transgenic fish combined with the GFP to differentiate two close regions, such as using red fluorescence in the liver and green fluorescence in the pancreas. Transgenic organisms or cells are those whose genome (their set of DNA) has been modified by artificially introducing one or more foreign DNA sequences extracted from a different species. In this case, the fluorescent proteins are transfected into the zebrafish cells to modify its genetic makeup, so it obtains the same fluorescent properties. Introducing fluorescence in specific organs in zebrafish aids in the study of those organs by image analysis, which can be done qualitatively and quantitatively[138], as the fluorescence may emit several different wavelengths depending on the type of protein.

Biobide is an organisation that specialises in the described zebrafish services[139]. Founded in 2005, their main goal is to accelerate and minimise risks in the research and development process for various industries, including pharma, chemical, and cosmetic companies, mainly in the preclinical process[139]. When Biobide was established, a significant problem faced by pharmaceutical and biotechnological companies was the minimal marketing success regarding new drugs and blockbusters. This medication generates annual sales of over $1 billion. Using zebrafish models, they have specialised in a variety of services that they can provide. To list some, they can aid in target validation, disease models, toxicity assays and efficacy assays[139].

Bringing back mCherry, a specific example of its use in zebrafish would be the thyroid disruption transgenic line, which works in the thyroid gland to express red fluorescence under the thyroglobulin promoter capable of directing thyroid-specific transcription. A fluorescence signal is obtained to be assessed and compared to a control mean by image analysis[140]. The analysis and calculations are used to ascertain the test's potential and hazard profile[140]. This allows the measurement of the fluorescence intensity of the thyroid gland, which is related to thyroglobulin levels. This assay widely screens thyroid-disrupting chemicals to complement other environmental ecotoxicity assays. Ecotoxicology studies the effect of toxic chemicals and other pollutants' on human health and the environment. To avoid ethical concerns of testing on animals, alternative assays have been developed to replace the old animal models used to test contaminant impact on humans and the ecosystem[141]. The cost is another advantage of these replacement ecotoxicity models compared to using animals[141]. An example of these alternative models is daphnia – a simple and efficient-to-culture tiny planktonic freshwater crustacean. One notable feature of these creatures is their transparent internal organs, which allow easy observation without intervention.

On the other hand, zebrafish larvae are another cheap and quick method that minimises the need for a vast collection of animals and the maintenance cost[141]. Using zebrafish six days or less after fertilisation removes the need for whole-fish testing as zebrafish are fertilised outside the body[141]. They also reproduce about every ten days. Also, like the daphnia, the zebrafish is see-through, providing the same advantages. This means that researchers can monitor and modify the developing fish quickly, and as mentioned, specific tissues are often fluorescently dyed to make them more distinguishable.

Going back to the other applications mentioned, apart from humans and mice, interestingly, the zebrafish genome has the most fully sequenced genome to date, meaning that we have determined the entirety of the DNA sequence of zebrafish. This, coupled with surpassing 14,000 gene mutations in zebrafish embryos having been effectively tested, is precisely why they have been chosen by Biobide to be applied across a wide range of industries.

In human disease research, scientists can reproduce genetic mutations in zebrafish model organisms to study and determine the time and location of their occurrence and their continual progression. The viability of pharmaceutical treatments is determined by testing on zebrafish. There has been progress in understanding disease and treatment in cancer, cardiovascular, and muscular diseases. In particular, one significant threat to human health is cardiotoxicity, and this is also a leading cause of drug candidate withdrawal in the preclinical stages. Another advantage of using zebrafish is that they have a heart biology similar to that of humans. Using this to our advantage, researchers can introduce green fluorescence to zebrafish hearts to monitor the heartbeats and record a video of the flowing heart. With this visual aid, researchers can determine heart functional alterations like arrhythmia. The same concept can be applied to the liver with image analysis, allowing better visualisation and identification of the entire area of the liver and individual liver cells.

Zebrafish are frequently used in the FDA's National Centre for Toxicological Research in "predictive toxicology" research. Tests on zebrafish are conducted to determine the possible effects of several substances and drugs, including nicotine, on humans. Similarly, in agrochemicals, the behavioural changes and neurotoxicity linked to pesticides can be detected with the help of zebrafish. It can be concluded that the zebrafish model organisms can become a successful tool for research with applications to many industries while adding an element of efficient and ethical considerations.

Another practical example of fluorescent dyes and probes used in medicine is flow cytometry. Cytometry is the quantification of various physical or biochemical characteristics of a cell. Flow cytometry is a laser-based method for classifying these cell phenotypes and features[142]. It can measure various properties, including cell size, cell granularity, and intracellular proteins across multiple disciplines like immunology and cancer biology. Besides identifying cell characteristics, it can also be implemented for cell counting or sorting, finding biomarkers of abnormalities and detecting microorganisms[142]. On a cell's surface, surface receptors pass information into the cell by modifying cytoplasmic proteins, controlling several crucial cell functions such as growth and development. If this signalling network is disrupted, it often leads to disease of some sort. This is why understanding these processes in normal and abnormal conditions is needed.

Wallace H. Coulter, using the Coulter principle, holds the patent for the first flow cytometry device based on impedance. However, Mack Fulwyler invented the forerunner to modern-day flow cytometers in 1965. Then, in 1968, the technology was extended further with the first fluorescence-based version by Wolfgang Göhde, although absorption remained the preferred method over fluorescence. Now, the surface receptors or intracellular proteins are often transfected or stained by fluorescent proteins (including GFP), dyes or fluorescently tagged antibodies[142].

This technology works with three parts of the system[143]. First, the fluidic system involves suspending the individual blood cells from a blood sample in a liquid, typically buffered saline solution[143]. This liquid is referred to as the sheath fluid in the specific context of cytometry and is used to transport the cells to the detection site. It flows at a high pressure, achieving a laminar flow, meaning it flows smoothly without turbulence. Once this state is reached, the cells are injected into the liquid at high pressure. Therefore, the individual cells can continue to be delivered in a uniform single-file manner. This optics section contains either a single or multiple lasers to generate light scattering and activate fluorescence, which is then passed through several filters so different photodiodes or photomultipliers can measure different wavelengths[143]. Visible light scatter can be measured via forward scatter (FSC), which gives information on the cell size by measuring how much of the laser beam goes around the cell, especially if there is a control bead of known size to be compared to find the relative size[143]. Another direction is the side scatter (SSC), which is measured at 90° to identify the granularity of the cell[143]. This is related to the internal complexity of the cell. Different cells contain a different number of internal granules like proteins or organelles in the cytoplasm. Those with more particulates are considered more granular, and these will be measured to have a greater side scatter since more of the laser beam will bounce off the internal components. The standard detectors used in flow cytometry are photomultiplier tubes (PMT) due to their high sensitivity and low background noise[144]. However, solid-state detectors like avalanche photodiodes and silicon photodiodes also show future potential[144]. The final electronics system converts the measured signals into digital ones that can be read by the computer and represented in a plot[143]. The flow cytometer machine can classify the cells by type and colour. With this technique, around 10,000 cells can be analysed in under a minute[142].

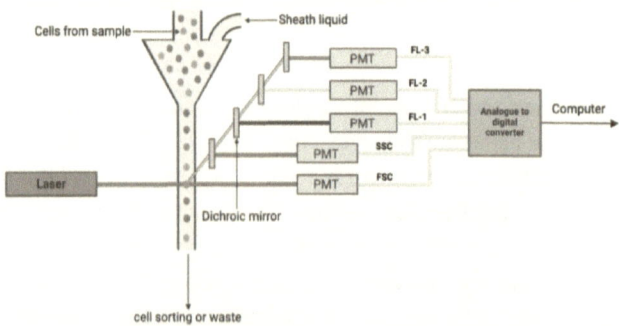

Diagram of possible cytometry structure. FL means it detects fluorescence.[10]

Multicolour flow cytometry refers to using different coloured fluorescent dyes when multiple types of cells or proteins are to be detected. On the other hand, "gating" refers to identifying and analysing only specific cell populations. Gating can be done through a specific type of flow cytometry called a cell sorter. The cell sorter uses a vibrating mechanism to oscillate the liquid containing the sample at high frequencies, causing individual drops to form[144]. Fixed parameters depending on the targeted population classify the drops as positively or negatively charged. Depending on this charge, these drops are passed through metal deflection plates to be separated and directed to their corresponding collection vessels, which can be plates, tubes, or slides[144].

Flow cytometry can also be combined with other photonics technology. For example, the combination of flow cytometry and fluorescence microscopy is an imaging flow cytometer (IFC). IFC can monitor the distribution of proteins in single cells, like in fluorescence microscopy, while also behaving as a flow cytometer to process vast numbers of cells; it allows for rapid analysis at both a single cell and population level[145]. This method is particularly useful in DNA damage and repair, intercellular interactions, and any other applications that require coordination between cellular position and expression of fluorescence in large cell populations[144]. Another example would be mass cytometers, which merge flow cytometry with time-of-flight mass spectrometry (TOF MS). TOF MS is a specific type of spectrometry that finds the mass-to-charge ratio of ions in the gas phase by measuring the speed at which they cover a known distance. In mass cytometers, fluorescently-tagged antibodies are replaced with heavy metal ion-tagged antibodies to be detected by the TOF MS[144]. Another feature is that the samples are destroyed during analysis, so cell sorting cannot be accomplished with this method[144]. Also, this method possesses only around 10% of the acquisition rate of a standard flow cytometer. Furthermore, if antibodies are unavailable, RNA Flow Cytometry, where RNA expressions and protein expression can be detected using fluorescent in situ hybridisation (FISH), can be employed as a helpful alternative[144].

The detection of different bacterial strands requires similar techniques. Identifying bacterial strains is critical for the theranostics of bacterial infections and the development of antibiotics[146]. Accurate identification of bacterial strains is a crucial step for effective theranostics of bacterial infections and is essential for discovering new antibiotics. As a side note, the word "theranostics" originates from the combination of the terms "therapeutics" and "diagnostics". This is another developing area of medicine where drugs or other treatment methods are merged to sequentially or concurrently identify and address health issues, saving time and money, potentially avoiding some of the unfavourable biological impacts that may occur when these tactics are used independently.

Returning to fluorescence, different fluorescent probes can identify different parts of different bacteria. For example, certain fluorescent probes can distinguish between enzymes present in different bacteria. Pathogenic bacteria are also classified as either gram-positive or gram-negative, depending on the composition of their cell surface[147]. Simplified, compared to gram-positive bacteria, which have a thicker cell wall, gram-negative bacteria contain an additional outer membrane with varying components and molecular composition[147]. These differing features allow the ability to detect and distinguish between bacteria. Needless to say, each type of bacteria has different effects on our bodies, which is why accurate, reliable, and efficient detection is necessary.

Countless organic fluorescent probes have been produced to overcome the flaws of existing detection methods. These newer probes can identify bacteria using an "off-on" fluorescent change, allowing real-time imaging in vitro and in vivo with quantitative analysis[148]. An example of an enzyme commonly expressed in most bacteria is nitroreductase, making it a suitable candidate to detect the presence of pathological bacteria. Some enzymes are specific to the strain. For example, E. coli, a common infection, is reported to highly express alkaline phosphatase[148]. Antibiotic-resistant bacteria, like Mycobacterium tuberculosis, exhibit high levels of β-lactamases, while Caspase-1 has been demonstrated to be activated when in bacterially infected human cells[148]. Based on our knowledge of specific enzymatic reactions present in different bacteria, we now have many options for fluorescent probes to selectively observe bacteria. Different fluorescent probes are tested on the enzymes to record what their respective absorption and emission wavelengths are[148].

The detection of mercury and chloride ions are two more instances of fluorescence detection to investigate the characteristics of bacteria. Mercury ions are highly hazardous substances that build up in tissues and organs, leading to a range of illnesses, such as renal failure and problems with the neurological system[149]. Since bacteria have been shown to convert Hg^{2+} to $Me-Hg^+$ in some instances[150] and analytes in Hg (II) sensors are typically contaminated by bacteria, using fluorescence to detect bacterial Hg^{2+} could aid in clarifying the mechanism and purpose of this conversion. According to some research, chloride is necessary for bacterial development and maintaining equilibrium between the concentrations of external salts[151]. Despite this, the significance of chloride for the bacterial cell's physiology remains mainly unclear. As a result, it is crucial to create fluorescent probes with legitimate chloride-binding sites. As indicated at the beginning, we can further enhance our detection and therapy procedures with greater knowledge of the characteristics of both general and specific bacteria. The technology's application in monitoring environmental water quality is one instance of its utility. The identification of bacteria is essential for preventing illness and infection since some bacteria, including E coli, can spread through polluted water.

Microscopes

But how are these bacteria detected? One straightforward way would be with the help of microscopes. As early as the 1660s, Dutchman Antonie van Leeuwenhoek grinded his own lenses to create a primitive version of a microscope which behaved more like a magnifying glass with one lens available. Nevertheless, these hand-made lenses could magnify an object up to 200 times its original size. With these lenses, Leeuwenhoek looked at living tissue, cells, fossils, and several other objects that had not ever been observed so closely before[152].

Meanwhile, Robert Hooke was making amazing discoveries with his microscope, and it was with his microscope that cells were first described. Over centuries, advancements in microscopy have allowed scientists to see clearer and more accurate images, expanding a whole new field of biology, allowing for the understanding of the activity of cells to details of cellular events. The current drive is to watch events in living organisms with ever more temporal and spatial resolution.

Fluorescent microscopy is one innovation that has granted us more power. How does this differ from conventional microscopy? Typically, a microscope uses visible light to illuminate through a lens and magnify the image of the sample[153]. However, if the sample of interest is fluorescent, a fluorescence microscope could be used instead. The mechanism behind this method is to adopt a much higher-intensity light source which excites the fluorescent species. This is where fluorophores come in, as the sample in question has to be fluorescent. They must be attached to the targeted features. Instead of the original light source, the excited fluorescent species generates a longer wavelength and a lower intensity light, resulting in the magnified image[153]. To put it briefly, the primary purpose of a fluorescence microscope is to provide illumination by excitation light for the specimen, after which the image's considerably weaker light is sorted out[154].

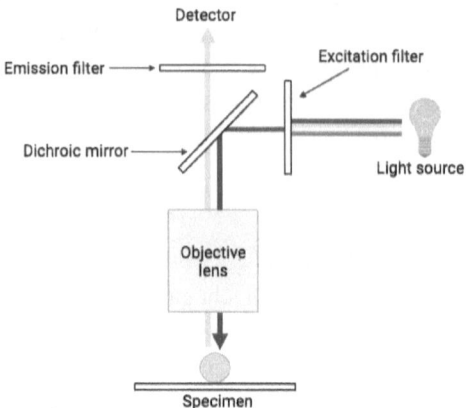

Diagram of the basic structure of a fluorescent microscope

Here is a more detailed explanation. After the sample of interest is labelled as necessary, it receives illumination from the high-energy source through the lens. As fluorescent materials only display fluorescent properties under specific wavelengths, the microscope needs a filter that selectively allows the required radiation from the energy source to pass through. The light allowed to pass gets absorbed by the fluorophores attached to the sample, leading to a longer, lower energy wavelength light emission. As explained earlier, electrons are stimulated to a higher energy level when the atoms in the specimen collide with the radiation[153]. They release light when they settle to a lower energy level. To become detectable, in other words, visible to the naked eye, a second filter is used to isolate the released fluorescence of the sample from the considerably brighter excitation light designed for that specific wavelength. This works because the light released has a longer wavelength and less energy than the light used for illumination due to some energy being lost in this process. This separation from the radiation from the surroundings allows the observer to view only what is fluorescing.

Another microscope to be compared is the electron microscope, another common type of microscope in which an electron beam is used instead of light to illuminate specimens. While this method can achieve resolutions and magnifications far higher than light microscopy[155], allowing for the significantly more detailed visualisation of microscopic structures and features, one crucial drawback of this technology is its limit in specimens. For specimens to be observed under an electron microscope, they are typically required to be treated and fixed in a manner that can damage the specimen's natural structure, rendering it unfit for studying living samples like cells and organisms. This is the leading advantage light microscopes, in general, have over electron microscopes.

Certainly, fluorescent microscopy is not limited to the field of biology. Digressing briefly, here are some examples of its uses in various fields. Mineralogical applications are a field that mainly benefits from fluorescence microscopy, and it frequently takes advantage of the technology for investigating minerals like coal[156]. In food chemistry, it is employed to evaluate a product's structural arrangement, presence, and spatial distribution of particular food ingredients[157]. There are some other miscellaneous fields where microscopy does not seem to have obvious applications. Take the textile industry, for instance, where fibre dimensions can be examined using microscopy. Fibre-based materials such as textiles and paper can be investigated with the help of epifluorescence microscopy. Fluorescence microscopy, which uses fluorescent dye, is also optimal for studying ceramic porosity[156]. This can also be applied to semiconductor research, such as how electron microscopes are used in inspection in manufacturing stages[156].

However, as with conventional microscopes, the most evident use of fluorescence microscopy would be in biomedical, biological and material sciences. Its high sensitivity and specificity help enable precise and detailed identification of cells and submicroscopic cellular elements. Some particles, like neurotransmitter amines, are invisible under standard microscopes, including adrenaline and serotonin[156]. Therefore, it can be seen that fluorescence microscopy has applications in both chemistry and histochemistry when it comes to biological tissue. Furthermore, fluorescence speckle microscopy is another application of fluorescence imaging. It is a technique that studies mobility and turnover rates, along with the replacement of ageing proteins as they degrade within the cell using fluorescence-labelled macromolecular assemblies[158].

But what is the need for fluorescent microscopy? One point is that the resolution is more limited since traditional light microscopy uses visible light[159]. Contrarily, fluorescence microscopy produces a picture that is far more accurate and detailed by using light generated by the fluorophores in the material[159]. Actually, for over a century, there existed a fundamental limitation not particular to fluorescence microscopy but rather optical microscopy in general that remained uncontested[160]. Its restricted spatial resolution results from light's wave nature, governed by its wavelength, usually between 300 and 800 nm. All of this changed in the 1990s of the previous century when Stefan Hell and associates discovered that it was possible to use the intrinsic quantum-mechanical characteristics of fluorescent molecules' fluorescence emission to overcome the diffraction limit[160]. Since then, superresolution microscopy has emerged as a new area of fluorescence microscopy[161]. Since then, a new area of fluorescence microscopy known as superresolution microscopy has developed. Even now, as the field continues to advance quickly, many publications have been published to support ever-more advanced techniques for superresolution microscopy[160]. That is not to say that light microscopy should be abandoned. In microscopy, conventional light microscopes continue to be widely used. They are beneficial in clinical settings for tasks like biopsy inspection because they are often easier to use and yield results quickly[159].

On the topic of superresolution fluorescence microscopy, numerous novel findings have been made that would have previously been impossible to attain with traditional optical microscopes. One interesting area of research is the application of superresolution fluorescence microscopy to 3D reconstruction of specimens, although that has been limited to thin samples. The reconstruction of three-dimensional (3D) structures from thick tissue samples using superresolution fluorescence microscopy can potentially transform biological research, particularly in brain connectomics and organism development[162]. The successful application of superresolution fluorescence microscopy for optical 3D reconstruction of dense biological material requires the resolution of two primary obstacles[162]. The first is that the single-molecule images are heavily contaminated by background noise. Strong auto-fluorescence from bulky samples and probe signals outside the imaging volume make up this noise[162]. The second issue is the photobleaching of probes occurring during imaging, particularly those situated outside the imaging volume and requiring imaging in subsequent procedures[162]. This refers to a cell's propensity to bleach when observed for an extended period of time using fluorescence microscopy.

Experiments and research have been done to improve 3D imaging techniques, such as using DNA-PAINT to avoid photobleaching[162]. PAINT stands for Point Accumulation in Nanoscale Topography and was initially developed as a dye-based strategy for localisation microscopy by exploiting fluorescent molecules that bind transiently to a target. The successive single-molecule signal produced by the quick binding and unbinding is known as fluorescence blinking, which can be utilised for single-molecule localisation[162]. Further explained, fluorescent blinking, also called fluorescence intermittency, is the random switching between an emitter's on and off states during its state of excitation. Also, fluorescence imaging has no temporal limit because photobleached probes are always replaced with new ones[162].

This method was applied to a slice of the mouse hippocampal area in a study to test the microscope's performance in thick tissue samples[162]. Presynaptic proteins were stained with one fluorescent dye colour, and postsynaptic proteins with a different colour in mouse brain tissues that were 100 μm thick. The findings demonstrated that thick tissue samples with synaptic gaps between 250 and 300 nm in size may be seen under a microscope[162]. They showed that, compared to some previous superresolution fluorescence microscopy imaging depths being only a few micrometres, they can now reconstruct entire 3D structures of objects with thicknesses up to 100 μm with the new microscope[162]. This discovery may open new doors for the widespread use of superresolution fluorescence microscopy methods in tissue imaging.

For example, the primary technique now accessible for high-resolution imaging of brain connections is serial slice electron microscopy. However, this method is highly tedious and prone to mistakes. It cannot distinguish between an excitatory and an inhibitory chemical synapse nor give distinct images of gap junctions[163]. Moreover, reconstructing a 3D neural connection map from 2D stacks of electron micrographs with greyscale images takes an enormous amount of time[163]. All these issues could be resolved with superresolution fluorescence microscopy, but their use in connectomics has been restricted because of the previously listed issues. Hopefully, better techniques, such as implementing DNA-PAINT, will improve the situation and give us a way to track the morphological changes of synapses during memory and learning. This will allow for improved diagnosis of disease and assessment of brain health as well as improved findings on the development of cognitive functions to better understand how our brain works and matures.

It is important to note that while DNA-PAINT offers better localisation accuracy and is not affected by photobleaching, one unfavourable aspect of the method is its poor imaging speed[162]. Further, newer techniques like FRET-PAINT have reportedly been able to overcome the issue of DNA-PAINT's poor imaging speed, increasing it by more than 40 times. However, greater advancements are anticipated with the introduction of shorter DNA probes and faster cameras. The future may see the effective fusion of confocal and FRET-PAINT microscopy, which would standardise the application of superresolution fluorescence microscopy for tissue imaging[162].

One specialised use for fluorescence microscopy is calcium imaging. Calcium imaging, as the name suggests, is a technique in microscopy that measures the calcium status of a media from tissues to a single cell optically to identify variations in the calcium stores' flux. Calcium indicators, fluorescent substances that exhibit fluorescence as a reaction to the binding of Ca2+ ions, are used in calcium imaging. What is the purpose of measuring calcium specifically? The idea behind this method is that intracellular calcium levels rise in response to neurons firing action potentials and that fluorescent compounds that attach to calcium can be used to identify this rise[164]. This means that researchers can potentially use calcium imaging to track neuronal activity over time and look into how networks expand or transform during learning.

Laser Speckle Contrast Imaging (LSCI)

Laser Speckle Contrast Imaging (LSCI), synonymous with Laser Speckle Imaging, is used to generate images of speckle patterns. Speckles are a form of noise that degrades the image quality and potentially complicates visual or digital analysis and interpretation. By firing a laser at a scattering (rough) surface, due to the random interference of coherent light, speckle patterns can be detected by photodetectors. These random interferences occur because light from various rough surface areas travels different optical path lengths to reach the image, causing the light waves to interact in different ways. As for the speckle size, this is determined completely by the aperture of the optical system used to observe the speckle pattern; it generally does not relate to the surface structure. The aperture refers to the opening in an optical device, such as a camera, that controls the amount of light that enters and reaches the image sensor or film. For example, if a person is directly observing the speckle pattern with their eyes, their pupil determines the speckle size. This becomes more important when a camera is used to capture the speckle patterns, as the setting of the aperture stop is crucial since it is used to control the exposure.

After the speckle pattern is captured, the local speckle contrast is calculated by measuring a small region, usually a 5x5 or 7x7 pixel grid from the image, and is defined as the ratio of the standard deviation to the mean average intensity of the speckles within that area[165]. The speckle contrast is a value between 0 and 1[165].

$$K = \frac{\sigma}{<I>}$$

K = speckle contrast, σ = the standard deviation of the intensity fluctuation and < I > = the mean intensity.

Examples of speckle patterns

One of the most common applications of laser speckle contrast imaging is in blood perfusion analysis. In such cases, when particles on the surface are in motion, there are fluctuations in the speckle pattern, affecting the intensities of the speckles and making them less clear. As a result, the speckle contrast in the grid decreases. From this idea, speckle contrast values can serve as an indicator of particle motion, providing valuable information about the movement of particles, including blood flow velocity[165]. Speckle contrast values range from 0 to 1, with 1 representing no particle motion and 0 signifying the fastest motion, resulting in the complete blur of all the speckles. Speckle contrast can also be expressed as a function of the camera exposure time (T) and correlation time (τ_c)[165].

$$K = \left[\frac{\tau_c}{2T} \left\{ 1 - \exp\left(-\frac{2T}{\tau_c}\right) \right\} \right]^{\frac{1}{2}}$$

The mathematical function exp(x) is equivalent to e (Euler's number) to the power of x. Camera exposure time refers to the duration the camera sensor is exposed to laser light to capture a single speckle pattern. Looking at the equation, it can be observed that shorter exposure times lead to speckle contrast values closer to 1, so less blurring of the images and vice versa[166]. The optimal camera exposure time has to be determined for each scenario as the duration is a trade-off between sensitivity and noise; longer exposure times have been found to have higher sensitivity but are more prone to speckle contrast noise[166]. Next, correlation time is the time taken for there to be a significant measurable change in the speckle contrast value, which, in the scenario of blood measurement, would indicate a change in blood flow. Depending on the context, different thresholds can be set to determine what counts as "significant change". There is also a formula for the correlation time, which is inversely proportional to the scattering particles' speed[166].

Before its increasingly widespread applications in medicine, Laser Doppler flowmetry was the norm. Laser Doppler flowmetry is based on the Doppler shift. This is the change in frequency or wavelength experienced by all waves, including light, after being reflected by moving objects like red blood cells[167]. A beam of low-power laser light guided by a fibre-optic cable enters the target tissue[168]. When the coherent laser light hits the moving blood cells contained in this volume, it bounces off and shifts in frequency due to the Doppler effect. The surrounding tissue also reflects the light but without any frequency shift. As a result, the illuminated volume consists of a blend of unshifted and Doppler-shifted light, with the magnitude and frequency of the shifted light being linked to the number and velocity of moving cells[167]. One main limitation of this method is that its perfusion signal is sensitive to movement's influence in the tissue or fibre-optic probe[167]. Due to this, many LDF studies have to be restricted to a single spatial location, allowing extremely high temporal resolution[167]. Spatial information can be obtained using LDF, but achieving this requires some method of scanning the beam and comes at the price of temporal resolution; scanning and data processing can take several minutes.

LSCI can overcome these limitations as it is a full-field approach using conventional CCD cameras and lasers[167]. It can create a map of velocity in a single shot, allowing it to achieve high spatial and temporal resolution simultaneously. This phenomenon was again brought about by the invention of the laser in 1960. Although "speckles" became more popular, this effect was initially called "granularity."[169] As mentioned, speckle is technically a type of noise, so when the laser first arrived, these random interference patterns were simply seen as a nuisance. When laser light was applied, such as in holography, it negatively impacted resolution, so much research was done to reduce speckles in images created using laser light[169]. In fact, there are even machine learning algorithms programmed to carry out this task of reducing speckle contrast noise. However, researchers soon began to study speckles to make use of the phenomenon for practical applications.

In 1980, the optics group at the University of Essen, Germany, headed by A.F. Fercher, aimed to develop non-invasive techniques for diagnosing eye conditions[169].

One of their projects involved measuring retinal blood flow, which helps identify various eye-related issues. At the time, the standard method involved injecting a fluorescent dye into a patient's bloodstream and waiting for it to appear in the blood vessels of the retina to identify issues like blocked blood vessels[169]. This is quite an invasive diagnostic method, and a non-invasive technique was thought to benefit both the patient since they don't have to be injected with potentially dangerous chemicals and the ophthalmologist since their examination time window would not be limited by the dye technique[169]. Thus, single-exposure speckle photography was created. The speckle pattern in an area of blood flow will be blurred to a degree correlated with the flow velocity and the exposure time compared to the speckle pattern in a motionless area, which will remain high in contrast[169]. This allows the velocity distributions to be represented and mapped as variations in speckle contrast[169].

When the proposed technique was applied to the monitoring of retinal blood flow, although the quality was subpar, it demonstrated the capability of spatial filtering in visualising retinal blood vessels. Even though single-exposure speckle photography had shown to be feasible, its appeal to researchers and clinicians was diminished by the fact that it required two steps to process the image, place the resulting transparency in the spatial filtering setup, and then take another picture[169]. The strategy was abandoned along with the Essen group disbanding in the middle of the 1980s.

Luckily, when the 1990s came, technology had improved enough for the technique to be reconsidered. The main point was to eliminate the photography stage so that the contrast could be measured directly, as this was the key disadvantage of the original method[169]. This new development required an adjustment in the name as it no longer involved photography. This is where the term Laser Speckle Contrast Analysis (LASCA) came from, although this term is more commonly replaced by LSCI or LSI[169]. The general setup was basically the same; the object of interest is illuminated by laser light to be imaged by a camera, typically a CCD camera. The image could then be captured by a frame grabber, which is an electronic device that "grabs" individual still frames from analogue or digital video signals to convert them into digital images[169]. It is often used in computer vision applications. Next, this data is passed to a PC to be processed by custom software.

The Kingston Group, who started this proposal, broke up in 1999, and so their work ceased[169]. This, however, did not lead to a halt in studies on the topic. The main areas for development included optimising the exposure time, reducing noise, and improving the original LASCA theory. The improvement in image quality meant the technology was being increasingly implemented. It is important to note that its applications are not restricted to medical fields. Examples of non-medical applications include tracking the velocity of vehicles on frozen roads and monitoring extremely slow-paced events like watching paint dry[170].

That being said, the medical field has been its primary field of application. Much research is being carried out on using the method for observing cerebral blood flow in animal tests. Microcirculation investigations are another common application. In general, the retina, skin, and brain are among the tissues that LSCI is particularly suited for since the targeted microvasculature is usually located near the surface, and LSCI does not perform as well in deep tissue due to the restrictions of the penetration of laser light. One of the first uses of LSCI full-field monitoring was in skin perfusion[171]. Initially, skin was widely used for in vivo test samples due to its convenience. That being said, imaging skin perfusion is challenging because most blood vessels are hidden beneath a layer of tissue with limited vascularisation, making it hard to track individual vessel flow[171]. However, LSCI can still measure overall capillary bed perfusion. Choi et al. have recently leveraged this approach to show that LSCI can be effectively paired with laser therapy to offer feedback during port wine stain (PWS) treatment[172, 173]. PWS is a type of birthmark or skin discolouration caused by a vascular anomaly. Although these are rarely indicators of diseases, if left untreated, the mark will continue to darken. During laser therapy for PWS birthmarks, pulsed visible laser light targets vascular lesions. Despite laser therapy being the primary treatment for PWS, some regions of residual perfusion may persist after treatment. LSCI can offer real-time feedback during treatment by visualising the skin's perfusion distribution before and after laser therapy.

Another one of the earliest applications of LSCI was the visualisation and analysis of retinal blood flow in both humans and animals. There are several reasons why a significant effort has been put into blood flow imaging techniques for the eye. For example, the retina is relatively easy to access, and blood flow is a crucial indicator of various ophthalmological conditions[171]. Research in the retina has explored the impact of different pharmacological agents on blood flow and examined flow patterns around the optic nerve head[171]. However, despite the numerous studies utilising speckle techniques to measure retinal blood flow, there is a lack of reported images of retinal blood flow. Most research has focused on presenting average blood flow values derived from images without providing spatial maps of the blood flow patterns[171]. This omission may be attributed to the limited spatial resolution of earlier instruments, which were restricted by small camera sensors (100×100 pixels) and, therefore, unable to capture detailed spatial maps[171].

Moving onto the first application mentioned above, when applying this technique to the brain, to enhance image quality, a section of the skull has to be either thinned or taken out and coated with saline or mineral oil, which helps to reduce the distorting results from static scattering elements[171]. The changes in blood flow that occur during brain activation are crucial in approximating oxygen consumption. From earlier discussions, functional magnetic resonance imaging (fMRI) is a standard method for investigating brain responses to different stimuli[171]. BOLD (blood oxygen level dependent) fMRI detects variations in blood oxygen, especially deoxyhemoglobin. Consequently, BOLD fMRI can sense and measure the hemodynamic responses occurring in the brain[171]. Hemodynamics simply refers to how your blood flows through your blood vessels.

Optical techniques such as hemodynamic responses have also been widely used for imaging. Intrinsic signal optical imaging (ISOI) is an especially common method. ISOI works by measuring cortical reflectance change on the principle that active brain tissue areas reflect less light than inactive regions when directly illuminated[171]. This means that the darker the brain area, the more active it is. One reason behind this is the differing absorption properties of oxygenated and deoxygenated haemoglobin. Others include changes in blood volume, and active neurons exhibit increased light scattering from swelling and movement of ions. This label-free and minimally invasive procedure maintains high spatial and temporal resolution, primarily achieved through techniques involving labels or markers. This method has helped provide valuable insights into the brain's functionality by creating functional maps and visualising the changes in cortical reflectance[171]. Through mapping the changes, researchers can pinpoint the brain regions that activate in response to specific stimuli or tasks to observe which areas are involved with different activities and possibly examine how different regions communicate with each other. Temporal dynamics can also be provided by ISOI so that researchers can observe how quickly the changes occur. Drug testing is a good application for this technology because it monitors changes in blood flow and oxygenation levels and allows researchers to evaluate how different substances or treatments influence brain activity.

These studies have mainly been based on qualitative mapping, and while valuable information on multiple aspects of cortical function can be obtained from this technique, the studies had yet to have been unable to extract quantitative spatial data on haemoglobin oxygenation, blood volume, flow, and metabolic parameters that underpin the observed signals[171]. While they may be able to provide relative measurements, they cannot offer absolute quantification of these values.

On the other hand, LSCI can also record the spatiotemporal changes in cerebral blood flow (CBF) following functional brain activation[171]. To measure the relative changes in cerebral blood flow (CBF) in response to a stimulus, speckle contrast images are taken at multiple time points during the stimulation, then averaged for each time point. After that, for every image, the speckle contrast values at each pixel are converted to speckle correlation decay times (τc)[171]. Finally, the ratios of the reciprocal of the decay times are calculated to indicate the relative changes in CBF[171].

As the CBF reaction to functional activation is just one part of the overall hemodynamic response, several research groups have integrated LSCI with other optical methods to simultaneously measure CBF along with other important hemodynamic and metabolic factors affected by functional activation[171].

Since LSCI can use any wavelength in the red or near-infrared regions, LSCI can simultaneously be performed alongside other techniques. Some examples include multispectral reflectance imaging (MSRI) for blood volume and haemoglobin oxygenation, measuring the partial pressure of oxygen through phosphorescence quenching, and fluorescence imaging of enzymes like flavoproteins and NADH[171]. There are quite a few new terms to break down. Starting with MSRI, this enables image data from specific ranges of wavelengths of the electromagnetic spectrum to be captured. This can be achieved using filters to isolate the desired wavelengths or specialised instruments that detect specific wavelengths. These wavelengths, such as infrared and ultraviolet, can extend beyond the visible range. This means MSRI allows a more comprehensive extraction of information than the human eye, which has limited capabilities with its visible receptors. Different combinations of spectral bands are used depending on the purpose. They are typically still represented with RGB channels. True colour refers to when only RGB channels are used and mapped to their corresponding colours[174]. Since it simply uses regular colours, this type of photograph is suitable for examining man-made objects and easy for beginner analysts to grasp. Other combinations include green-red-infrared, used for detecting vegetation since it is highly reflective in the near-infrared[174]. In this combination, the near IR replaces the usual blue channel so that any reflection of infrared shows as blue.

To combine this with LSCI, a second camera and light source must be added to the original LSCI setup. Incoherent light ranging from wavelengths usually 550-650 nm successively illuminates the cortex for MRSI, while speckle imaging uses laser light of longer wavelengths from 700-800 nm[171]. The light reflected from each source is split into distinct colours and captured by two cameras, which record both the speckle pattern and the multispectral signals. Next, for MRSI, the spectral images are transformed into maps of oxygenated haemoglobin (HbO) and deoxygenated haemoglobin (HbR) changes at each pixel, visually representing haemoglobin fluctuations[171]. The LSCI are processed using the previously explained method.

The other optical method of phosphorescence quenching measurements of partial pressure of oxygen (PaO2) allows us to non-invasively measure oxygen levels in the microvasculature of tissue[171]. PaO2 is a type of blood gas tension, which is the partial pressure of gases found in the blood. In simpler terms, this is the pressure exerted by an individual gas within a mixture[175]. The sum of every partial pressure in a mixture makes up the total pressure of a gas. Common gas tensions measured include the partial pressure of oxygen, carbon dioxide, and carbon monoxide. PaO2 reflects the effectiveness of oxygen transfer from the lungs into the bloodstream[175]. This pressure propels oxygen to diffuse across the alveolar membranes, through the walls of pulmonary capillaries, and into the bloodstream, where it's carried by red blood cells to peripheral tissues throughout the body[175]. That is why it is a vital parameter to measure and take into account. Phosphorescence quenching is the process in which the phosphorescence intensity of a given substance is decreased. Phosphorescence is similar to fluorescence but differs in the duration between absorbance and photon emissions. Fluorescence occurs when a substance absorbs light and emits a lower energy photon almost immediately, whereas phosphorescence happens over a longer duration[176]. This is because, unlike fluorescence, there is an extra energy level between the ground level and the excited level, called a metastable level[176]. It is also called an electron trap because moving from the metastable level to other levels is extremely unlikely, which is why it's called a forbidden transition[176]. The electron stays in the metastable state until it undergoes this forbidden transition or gets excited again to return to the transition level[176]. To produce radiation, two possible ways this excitation could occur are thermoluminescence, which is the agitation of nearby atoms or molecules through heat, or optical stimulation[176]. The duration of the metastable state determines how long phosphorescence lasts. The length of time can range from approximately 0.001 seconds to years or even decades, depending on the specific conditions[176]. Phosphorescence quenching measurements are based on the principle that oxygen can quench the emissive triplet states of exogenous probes dissolved in biological fluids.[177] The degree of phosphorescence quenching is directly proportional to the oxygen concentration, enabling the measurement of PaO2 with high precision in both time and space across three dimensions[177]. This technique is essential for studying oxygen behaviour in various biological events and conditions.

Another early application of LSCI relating to the brain was spreading depolarisation (SD). SD, also known as cortical spreading depression, is a slow-moving wave of reduced electrical potential (negative potential shift) that spreads across the brain's cortex at 2 to 5 millilitres per minute[171]. This phenomenon was first observed in rabbits in 1944 and is linked to increased metabolism, hemodynamic changes, and significant ionic shifts. Its well-established role in strokes, migraines as well as subarachnoid haemorrhage has been studied for over half a century[171]. Subarachnoid haemorrhage refers to bleeding in the space located between the brain and the thin layers of tissue that surround and shield it. However, the physiologic changes taking place during SD still require further understanding. Notably, more research is needed on the influence of cerebral microcirculation on the effects of SD on the brain. In animal studies, spreading depression (SD) can be triggered in normal brain tissue by applying potassium chloride topically or causing a localised injury, such as a pinprick[171]. Once initiated, SD moves across the cortex, causing significant transient shifts in direct current potential. As cells need more energy to restore their normal state, SD is usually followed by a substantial transient rise in blood flow (hyperemia) due to the increased metabolism. Laser Doppler has been proven effective in measuring and timing the transient hyperemia during SD[171]. However, methods like laser Doppler and MRI-based techniques lacked the spatial resolution to pinpoint blood flow in individual blood vessels. Instead, when LSCI was used, a temporary, yet delayed, surge in blood flow in the blood vessels was found in the dura vessels, which lie directly above the cortex[171]. Bolay et al. used imaging to track changes in blood flow within the middle meningeal artery, allowing them to identify the specific nerve pathway that connects the initial event of SD, which triggers a headache, to the subsequent development of the headache[178].

Optical coherence tomography angiography (OCTA) and optical coherence tomography (OCT)

Other non-invasive in-vivo imaging examinations include optical coherence tomography angiography (OCTA) and optical coherence tomography (OCT)[179]. Using light waves and interferometry, they can obtain precise cross-sectional images of your retina within 10 to 15 microns. When combined with catheters and endoscopes, OCT can serve as an optical biopsy, allowing for high-resolution intraluminal imaging of organ systems[180].

An ophthalmologist, a specialist in eye diseases, may view every unique layer of the retina with OCT. An ophthalmologist can map and gauge their thickness to aid in diagnosis. They also direct retinal disease treatments such as diabetic eye disease and age-related macular degeneration, as well as glaucoma. OCTA captures blood vessel images within and beneath the retina. OCTA is similar to fluorescein angiography, an eye test using a special dye and camera that looks at blood flow in the retina and choroid, the two layers in the back of the eye, except it is a faster test that doesn't require a dye. Since its initial introduction in 1991 by Huang and colleagues[181], OCT has been a widely used diagnostic technique in ophthalmology because of the transparency of the eye. In fact, deploying OCT for eye imaging has had one of the largest clinical impacts on ophthalmology. Soon after its discovery in 1993, the first human optic disc and macula in vivo tomograms were demonstrated. In 1996, the technique was commercially commercialised for ocular diagnostics after being transferred to industry. Furthermore, it has been applied extensively outside ophthalmology to image-specific non-transparent tissues.

To expand on that last point, OCT is the equivalent of ultrasound imaging, with the apparent exception that light is used rather than sound. However, picture resolution of 1 to 15 μm can be achieved, which is actually a factor of one to two greater than that of traditional ultrasound. Recent advances in OCT technology have allowed for the imaging of non-transparent materials., opening up new applications for OCT across various medical disciplines. Like other imaging technologies, tissue scattering and absorption-induced optical attenuation limit the imaging depth. In most tissues, however, imaging can be accomplished down to a 2 to 3-mm depth, the same scale as conventional biopsies and histology routine images. Despite the shallower imaging depth, OCT's resolution is over 10 to 100 times sharper than typical clinical ultrasound.

Before the OCT exam, the patient may require dilating eye drops to be applied, enlarging the pupil and facilitating smoother retinal examination[179]. The patient will then face the OCT machine while seated with their head resting on support to remain still for the equipment to scan their eye without coming in contact[179]. It takes 5 to 10 minutes to scan. Any condition that hinders light from entering the eye renders the scanner useless[179]. One possible observation would be a pearl necklace sign, composed of hyperreflective dots arranged in a contiguous ring in one of the layers of the retina[182]. It is usually seen in exudative macular diseases when abnormal blood vessels grow behind the furthest part of the retina.

Surface plasmon resonance (SPR)

There is another method of sensing that does not require labelling as with fluorescence technology. This would be using the concept of surface plasmon resonance (SPR). These types of sensors have come a long way in the past two decades, and they are now widely used in fields like chemical detection, material characterisation, and biosensing. Its increasing popularity in the biosensing field can be attributed to its exceptional specificity and sensitivity in identifying biomolecular interactions. Firstly, plasma is the fourth state of matter, along with gases, liquids and solids. It comprises ions, neutral atoms and electrons as a partially-ionised gas[183]. It can occur when heat is added to a neutral gas to a point where part of the electrons from the atom or molecules are liberated, causing it to change state and become a plasma[183]. They are generally composed of a mixture of negatively charged electrons and positively charged ions. Now, plasma oscillation is made up of plasmons, just as optical oscillations, also known as light, are made up of photons. Plasmons can be more precisely defined as the quantum oscillation of the free electron clouds about the stationary ions, which may sound similar to the structure of metals. As a result, when light strikes a metal conductor's surface at the correct wavelength and angle, plasmons may also form. Surface plasmons are specifically restricted to surfaces and have a strong light-interacting force. Though this holds for any chemical element, it is necessary to remember that although metals have plasmons, they cannot be classified as plasma until heated to a high enough temperature for the electrons to separate from the atoms and create a cloud surrounding the ions.

Plasmonic-based sensors have been developed for over four decades, and various products have been made available for purchase. They have a few advantages over conventional sensors, including the ability to monitor in real-time, label-free detection, high sensitivity, quick response times, and excellent reusability. Plasmon-enhanced fluorescence, also known as surface-enhanced fluorescence spectroscopy (SEF), is a process in which a fluorophore material's fluorescence intensity is increased with the help of a plasmonic nanomaterial. This is often accomplished by placing the fluorophore close to a metallic nanoparticle. Consequently, an augmented electric field is experienced by the fluorophore, leading to an amplified emission. In POC devices, SEF is frequently used to sense a whole range of analytes.

Photoacoustic imaging

Lastly, an ultrasound is a kind of imaging exam that creates images of your body's organs, tissues, and other structures using sound waves. We can extend its functionalities with the technique of photoacoustic imaging, which relies on the photoacoustic effect, meaning it combines optical absorption and acoustic detection[184]. When tissue is illuminated by pulsed laser light, substances with optical absorption properties, such as haemoglobin, undergo thermoelastic expansion, generating an acoustic pressure wave[184]. This wave can then be detected with an ultrasound transducer, and the initial distribution of light absorption is reconstructed to create an image. If more than one laser excitation wavelength is used, this is called spectroscopic PA imaging[184]. As tissues have distinct spectral signatures, spectroscopic analysis can be carried out for tissue characterization and quantification. Combining optical and acoustic signals enables dual-modal imaging with ultrasound and photoacoustic imaging[184]. The shared hardware is a primary advantage of this as simply by adding an additional optical source, currently installed ultrasound systems can be upgraded to PA/US systems rather than creating a new system[184]. This also means that ultrasound imaging can be incorporated seamlessly with photoacoustic imaging. Secondly, PA imaging enables better imaging resolution in depths going beyond the optical scattering limit since acoustic waves scatter significantly less than light in relation to tissues[184].

Similar to fluorescence imaging, different agents exhibit varying degrees of light absorption at different wavelengths, leading to distinct photoacoustic absorption spectra. It is possible to differentiate between various molecular agents by utilising the different excitation spectra. For instance, atherosclerotic lesions, or anomalies in the arteries, are frequently mapped using wavelengths of 1210 nm and 1720 nm to identify lipids. Additionally, intrinsic and extrinsic chromophores exist akin to fluorophores, which are elements of a molecule responsible for its colour. In biological systems, these chromophores undergo molecular changes in response to light. Melanin, haemoglobin, water, and lipids are examples of such chromophores. Since haemoglobin is a main component of blood, measuring its properties brings many crucial benefits. By taking advantage of PA modality, we enable the quantification of functional characteristics of blood vessels, including oxygen saturation and blood flow[184]. Several diseases can be detected early on by vascular anomalies, making the ability of photoacoustic imaging to give data and analyses on blood vessels highly valuable for a large variety of applications in the clinical setting.

This ability is especially useful in oncology. For example, the ability to measure blood oxygen saturation is key for cancer detection and staging as it allows us to use the differences in oxyhemoglobin and deoxyhemoglobin or in other words, oxygenated versus deoxygenated blood[184]. This is because most tumours experience hypoxia while growing, when insufficient oxygen is provided to the tissue. Another key feature is tumour angiogenesis, where new blood vessels form surrounding the tumour region[184]. By using a dual-modal system, the PA can image the oxygen saturation while the ultrasound can image the morphological characteristics of the tissue, like shapes and boundaries. Currently, MRI and CT scans are commonly employed clinically for cancer imaging. However, the distinctive benefits offered by PA/US dual-modal systems, including non-invasiveness of radiation, have contributed to their promising advancement in this field[184]. Not only can this technique be used for screening and diagnosis, but it can also be used for preclinical research and treatment response monitoring. In contrast, nanoparticles and chemical dyes that can be incorporated into the body are referred to as exogenous chromophores. These exogenous agents enhance the contrast of photoacoustic imaging and can be used in molecular imaging as targeting agents, further expanding the capabilities and applications of the technique[184].

There are various types of applications. One type of photoacoustic tomography is known as photoacoustic computed tomography. This is where the waves generated from the acoustic source are reconstructed[184]. An unfocused ultrasonic transducer array, a collection of uniformly pulsing elements used to guide soundwaves in a certain direction, generates an image by scanning over the source[184]. There is also such a technology as photoacoustic microscopy, a hybrid method used for in vivo imaging. It acoustically forms an image of the sample using the photoacoustic effect[184]. Photoacoustic microscopy aligns the focus of its acoustic detection and optical illumination to ensure maximum detection sensitivity when detecting the pressure change caused by the heat generated by the photons absorbed by the sample[185]. Similarly, photoacoustic endoscopy is used for internal organ imaging.

Biolasers

Circling back to one of the fundamental components in photonics, lasers. As mentioned, different lasers utilise different gain mediums, giving them distinct properties and purposes. Lasers are a major milestone and one of the most significant inventions from the 20th century. Still, so far, they have relied on artificial or engineered optical gain materials. On the other hand, biological lasers, or biolasers for short, are those which make use of biological entities, such as cells, tissues, and viruses as the gain medium. By using naturally derived biomaterials, these lasers have biocompatibility and can serve as ultra-sensitive instruments for detecting and imaging biological systems, as their optical output is closely tied to the structures and functions of living organisms[186].

This is where green fluorescent proteins come back, but with a different use. In 2011, Gather and Yun demonstrated the concept of constructing a "living" laser by incorporating cells containing GFP between a pair of dielectric mirrors[187, 188]. Instead of being used as a biomarker for detection, like in fluorescent microscopy, it can be integrated into the detection device itself. Gather and Yun achieved this by extracting cells from a human kidney and introducing the DNA responsible for coding GFP[189]. Subsequently, the researchers positioned some of the GFP-producing cells between a pair of tightly spaced mirrors, only one cell apart[189]. To initiate lasing, the GFP within the cells required excitation from an additional laser, which emitted low-energy blue light pulses, approximately one nanojoule. Ordinarily, blue light would cause the GFP within the cells to fluoresce, randomly emitting light in all directions, as we explained earlier[189]. However, within the confined cavity, the light underwent multiple reflections, amplifying the emission from the GFP and generating a coherent green beam just as other lasers with other gain mediums[189]. An intriguing advantage lies in the continuous production of GFP by biological cells, unlike the amplifying materials in conventional lasers that degrade over time. Consequently, this technique could lead to "self-healing lasers" that do not require replacement[189].

Remarkably, the researchers established the feasibility of creating a laser cavity using such an unconventional approach without harming the cells involved in the lasing process. As different cells possess unique structures, they generate laser beams with distinct characteristics, potentially enabling the acquisition of structural or functional information about the cells through the lasing process. Researchers have explored various biomaterials in these lasers, ranging from proteins and DNA at the molecular level to cells and tissues at the macroscale. Other materials like quantum dots and nanowires have also been incorporated into the system. This technique is promising for targeting diseased cells by activating focused light once the laser-producing cells reach the target tissue.

An example of an event indicating the significance of this emerging technology is in 2018, where the first Gordon Research Conference entitled "Lasers in Micro, Nano and Bio Systems" was held. The Gordon Research Conferences, originating in 1931, are a series of worldwide scientific meetings addressing cutting-edge research in the physical, chemical, and biological sciences and associated technologies arranged by an identically named non profit organisation[190]. Having expanded to almost 200 conferences per year, the conferences were initially hosted in the US but since 1990 also in Europe and Asia, showing its global significance.

As explained, fluorescence from dyes and fluorescent proteins has widely been used in analysing biomolecules. A sensing signal is produced when the chemical interactions connected to the fluorescence probes alter the properties of the fluorescence emission, such as its intensity and spectrum[191]. Nevertheless, the situation when the signal is overly faint and lost in the surrounding noise is a problem we frequently encounter[191]. However, by putting the probe molecules within a laser cavity, one may think of a whole different strategy using stimulated emission from the molecules as the basis rather than fluorescence[191]. This adaptation can potentially enhance the signal-generating process, leading to enhanced biomolecule detection and analysis sensitivity. The optical feedback that the laser cavity provides allows for the amplification of the signal in the laser, in contrast to biological amplification mechanisms like polymerase chain reaction, which simply double the amount of molecules to increase the sensing signal[191].

All this builds up to the creation of the optofluidic bio-laser, a novel type of laser that uses biological or biochemical molecules as the gain medium. The fluidic environment in which the sensing molecules are found exists in a living cell or, more widely, interstitial tissue and microfluidic devices[191]. The optofluidic bio-laser, which outperforms or supplements traditional fluorescence-based detection, has seen a rapid increase in biosensing applications since its introduction less than ten years ago[191]. Since bio-events frequently take place in watery environments, optofluidic dye lasers—which combine microcavities, microfluidic channels, and gain medium in liquid—have shown to be a valuable tool in biolaser research[192]. In fact, the development of biolasers has benefited immensely from and been inspired by the research in optofluidic dye lasers.

An early instance of this technology was a liquid dye laser in 1970, in which the substrate was a biocompatible "edible" Jell-O substance that had dissolved fluorescein dye[193]. The configuration of mirrors or other optical components encircling the gain medium is known as the laser cavity[194]. However, in this instance, vigorous pumping of the dye solution in the gelatin results in enhanced spontaneous emission since there is no particular cavity structure present. A cavity's optical feedback is necessary for laser oscillation. Several parameters, including the absorption cross-section, the concentration of stimulated fluorophores, and the emission cross-section, control laser emission properties. The underlying biological and biochemical processes can be uncovered by tracking variations in the laser's output properties (e.g. spectrum, intensity, and threshold), as these parameters might change in response to particular biomolecular interactions in an optofluidic bio-laser[191]. Since fluorescence microscopy can also detect low analyte concentrations, the main benefit of laser-based detection is its ability to identify minuscule signals that are otherwise difficult to distinguish from the relevant biological process and biochemical interaction[191]. Because of the enhanced signal-to-noise ratio, compared to fluorescence-based detection, the laser-based detection has a sensing dynamic range that is one to three orders of magnitude larger.

Laser-based detection is fundamentally different to fluorescent-based detection. These operate based on a laser light beam emitted by a sensor, which then receives a reflection from an object passing by and interprets it as a detection. It has several advantages over its fluorescent counterpart. Imagine a test tube with fluorescent molecules; a broad spectrum of fluorescence is released in all directions. Some fluorescence is contained in the cavity that mirrors define when the identical sample is put between two mirrors. In the test tube, stimulated emission amplifies this fluorescence every time light travels across the gain medium[191]. The cavity's subsequent emission has spatial, spectral, and temporal properties that are very different from those found in equivalent fluorescent detection devices in many ways[191]. Since the cavity controls the direction in which the laser output is produced, the output intensity of the laser is typically far higher than that of the omnidirectional fluorescence light[191]. Furthermore, there is a noticeable threshold behaviour in the output intensity, and the spectrum is multiple orders of magnitude narrower[191].

As seen in the span of the last ten years, we have witnessed a transition in microbiology from bulk to single-cell studies, driven by the abundant developments in the field of microbiology. Microfluidic devices now allow for accurate control of the micro-environment within single cells in long-term studies, and advanced microscopic techniques like time-lapse microscopy or the ones mentioned above have greatly expanded spatial and temporal resolution, enabling quantitative and dynamic measurements on individual cells. Having successfully cultured and visualised individual cells, the next task is to manipulate and manage them so that, for example, the reaction of each cell upon cell-to-cell interactions or contact with surfaces may be studied[195].

Optical tweezers and stretchers

This leads to another technology relating to lasers in medicine: optical tweezers. They were initially known as single-beam gradient force traps, and this scientific instrument is named as such because it makes use of an intensely concentrated laser beam in a similar fashion to tweezers to grasp and transport microscopic and submicroscopic objects such as atoms, droplets, and nanoparticles[196]. They have several applications, including in nanoengineering and nanochemistry, for studying and constructing materials made of individual molecules, quantum optomechanics, and quantum optics to investigate how single particles interact with light. In medicine and biology, optical tweezers are used to grasp just one single cell, from bacteria to sperm to blood, as well as molecules like DNA, for example[196].

The optical tweezer was granted the Nobel Prize for Physics in 2018, but its discovery came decades before. Arthur Ashkin was the scientist who first reported the observation of gradient forces and optical scattering on micron-sized particles in 1970[196]. Years later, in 1986, the method now widely known as optical tweezers was initially observed by Ashkin and his colleagues, who could hold small particles steady in three dimensions using a narrowly concentrated light beam[196].

Steven Chu, one of the authors of this groundbreaking paper, later employed optical tweezing as a part of his research on neutral atom cooling and trapping[196]. Together with William D. Phillips and Claude Cohen-Tannoudji, Chu was awarded the Physics Nobel Prize in 1997 for this work. Steven Chu explained in an interview how Ashkin first thought about optical tweezing as an atom-trapping technique. Chu used a magnetic gradient trap and a resonant laser light to extend Ashkin's methods of trapping bigger particles with diameters ranging from 10 to 10,000 nanometers to trap neutral atoms with a diameter of 0.1 nanometers[196].

In the late 1980s, Joseph M. Dziedzic and Arthur Ashkin also proved the technology's first use in the biological sciences by utilising it to trap a single Escherichia coli bacteria and a single tobacco mosaic virus[196]. Researchers like Steven Block, James Spudich, and Carlos Bustamante pioneered the application of optical trap force spectroscopy in characterising molecular-scale biological motors during the 1990s and beyond. Found throughout biology, these molecular motors are in charge of both internal cell motion and mechanical action. These will be explained more later. In general, these biophysicists have been able to study the dynamics and forces of nanoscale motors at the molecular level with the help of optical traps[196]. The stochastic character of these force-generating molecules has since been clarified by optical trap force spectroscopy[196].

Over the years, optical tweezers have been shown to be useful in plenty of other fields within biology as well. In synthetic biology, they are used to build artificial cell networks that resemble tissues and to join synthetic membranes to start biological reactions[196]. In genetic investigations and chromosomal dynamics research, they are also commonly used. In 2003, cell sorting was among the areas in which the methods of optical tweezers were utilised[196]. Cells may be sorted based on their inherent optical properties by forming a broad pattern of optical intensity throughout the sample region. This is a separate method of cell sorting compared to flow cytometry. Although it still uses a laminar flow to concentrate the targeted cells, using optical tweezers instead of metal deflection provides advantages like greater flexibility, higher recovery rate, and purity. A more specific example of an application would be researchers being able to trap and manipulate the proteins that whip a flagellum, the tail of a sperm cell, and measure the force of its swimming motion.

Moreover, the cytoskeleton, cell motility, and visco-elastic characteristics of biopolymers have all been studied with optical tweezers. Proposed in 2011 and experimentally proven in 2013, it is a bio-molecular assay demonstrating how ligand-coated nanoparticle clusters are recognised and trapped optically following target molecule-induced clustering. Anything from DNA strands, dielectric spheres, live cells, viruses, bacteria, organelles, and tiny metal particles has been captured using optical tweezers[197]. Applications range from organisation and confinement (such as cell sorting) to tracking movement (such as bacteria), application and measurement of small forces, and altering larger structures like cell membranes[198].

The fundamental idea underlying optical tweezers is the momentum transfer related to light bending[198]. The momentum of light is proportional to its energy and propagation direction[198]. Can light possess momentum if it has no mass? The formula of momentum is equal to mass times velocity is one that many would've heard of. However, in reality, momentum does not require mass. It only requires energy, and while light may not have mass, its photons carry energy. One example demonstrating that light has momentum would be a comet in space. Comet tails comprise ionised gas and gases that have evaporated from the comet's body. A common misconception is that these tails trail behind the comet when, in actuality, they always point in the direction opposite the Sun. When photons from the Sun scatter off these dust particles, they recoil, opposing the Sun. This indicates that photons travel further from the Sun with momentum and that, in collisions, this momentum is partially transferred to the dust particles[199]. However, since protons and electrons make up most of the cosmic radiation from the Sun, other particles affect the blue tail's gas molecules and atoms more than photon momentum[199].

Nevertheless, some dust particles are driven away from the Sun by light reflecting off them[199]. Additionally, photons interacting with atoms in the comet substance also form the ionised gas tails[199]. Usually, the tweezer beam is concentrated by passing it through an objective lens of a microscope. The oscillating electric field's amplitude varies rapidly near the beam waist, which is close to the narrowest point of the concentrated beam. The centre of the beam is the area with the highest electric field, where dielectric particles are drawn down the gradient[196]. Additionally, as the beam propagates, the laser light tends to exert force on the beam's constituent particles[196]. This is because of the principle of conservation of momentum, which means that photons are dispersed or absorbed by a small dielectric particle to give the dielectric particle momentum.

Using this idea of the particle being able to move in the beam, the research in molecular motors and the physical characteristics of DNA have been two of the primary applications of optical traps. Molecular motors include motor proteins. These proteins are bound to a polarised cytoskeletal filament, part of the network of interlinking protein filaments in the cytoplasm of all cells. They proceed steadily along it using the energy produced by repeated rounds of ATP hydrolysis. As a result, this can be simulated by using the beam's energy and momentum to move it. In both areas, a biological specimen is biochemically attached to a micron-sized glass or polystyrene bead that is then trapped. We have been given the capability to investigate different motor properties by affixing a molecular motor, such as RNA polymerase, to such a bead[198]. Some questions that can be answered include whether the motor moves in discrete steps, what size the steps are and what the maximum force output is[198]. Similarly, by attaching the beads to the ends of single pieces of DNA, experiments have measured the elasticity of the DNA as in its resistance to stretch, twist and bend, as well as the forces under which the DNA breaks or undergoes a phase transition.

In order to make optical tweezers more accessible to researchers on a tighter budget, researchers have endeavoured to reduce the size and complexity of these instruments[196]. The fact that a lens cannot concentrate a beam of light less than half its wavelength presents another major issue with optical tweezers[200]. This implies that the trapping will be inaccurate if the targeted particle is much smaller relative to the focal point. The focal size limit also restricts the maximum gradient force produced. Nanoparticles require a higher force to be trapped than relatively larger microscopic particles[201]. Hence, traditional optical tweezers must employ a very powerful laser to capture the smallest targets[201].

A few years back, applied physics researchers found a way to circumvent these issues by strengthening the trapping field by directing the laser onto a series of tiny gold disks[200]. The light causes the metal's surface electrons to become excited, leading to plasma oscillations — rapid waves of electromagnetic charge — which generate "hot spots" of stronger fields at the edges of the disk[201]. The small gold disks were arranged on a glass sheet according to the designs of other researchers, and the entire apparatus was immersed in water containing the target particle. Ken Crozier, an electrical engineering associate professor at Harvard, and his team, discovered that the biggest problem with those devices used in the tests was that the water boiled unless the laser intensity was maintained very low[201].

The Harvard team has found solutions to both issues by using silicon plated with copper and gold to replace the glass, with elevated gold pillars as the top surface[201]. These materials are much more thermally conductive than glass, so they act as a heat sink. The gold, copper, and silicon under the pillars act like the heat sink attached to the chip in your PC, drawing the heat away. Therefore, the new device reduced the water heating by about 100-fold with the hotspots produced at the top edges of the pillars; the team succeeded in trapping polystyrene balls as small as 110 nanometers.

From the smallest building blocks of matter to the basic units of life, laser traps have proven to be useful for manipulating various objects for nearly thirty years[202]. The fundamental idea of laser traps is that momentum is passed from the light source to the object, and that force is then applied to the object via Newton's second law[202]. These optical forces have been utilised just to trap objects for a while[202]. The one-beam gradient trap, also known as optical tweezers, had been the most common laser trap[202].

The optical stretcher, in contrast, uses a double-beam trap, which traps an object in the centre using two identical but opposing and slightly diverging laser beams. Like cells, the momentum transfer in extended objects mostly occurs at the surface[202]. Due to the symmetry of the two-beam trap geometry and the cancellation of all resultant surface forces, no force is exerted on the centre of gravity[202]. However, the forces on the surface stretch the item along the beam axis if it is sufficiently elastic[202]. Although this optical stretching initially seems illogical, there is a straightforward explanation for it[202]. Using the concept of light carrying momentum, consider a ray of light passing through a cube of optically denser material. As it enters the dielectric object, the light gains momentum so that the surface gains momentum in the opposite direction[202]. Light loses momentum as it leaves the dielectric object, causing the opposing surface to gain momentum in the direction the light propagates towards since the total momentum must always be conserved and equal[202]. The reflection of light on either surface also leads to momentum transfer on both surfaces in the direction of light propagation. The increase in the momentum of light within the cube accounts for a larger portion of the surface forces than this one[202].

In summary, the object tends to stretch since the two consequent surface forces on the front and backside are opposed. However, there is an asymmetry between the surface forces, leading to a total force that acts in the cube's centre[202]. If a second identical light beam enters the cube from the other side, the forces acting on its surface are additive, although no total force is acting on the cube[202]. This is one primary difference between optical traps and stretchers. The optical stretcher uses surface forces to stretch objects, as opposed to asymmetric trapping geometries, in which the trapping force involved in optical traps is the total force[202].

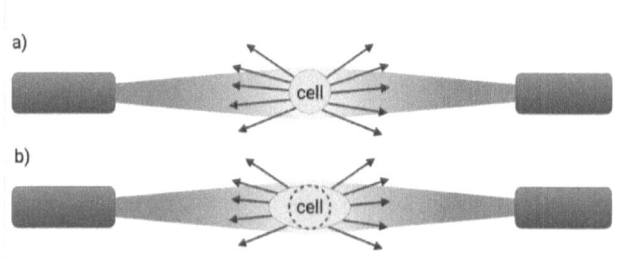

Diagram of optical stretcher. Two optical fibres to emit two lasers with the cell a) before and b) after stretching. The arrows represent the direction of the forces acting on the cell.

Since cell elasticity can be measured, different cell types or between normal and unhealthy cells can be differentiated by comparing this material property of cells. This has especially shown potential in cancer detection, as studies have been done on whether single malignant and precancerous cells can be detected. For example, although molecular biology has yielded a multitude of findings regarding cancer biology, it still lacks the ability to discern fundamental differences between benign primary and malignant tumours. Primary tumours are the first tumours that form, and these are separated into malignant tumours, which are those that can spread cancer cells throughout one's body through the blood or lymphatic system, versus benign tumours, which stay in their primary location without invading other sites of the body making them usually unproblematic.

As such, during the last 20 years, the role of cell mechanical properties in cellular processes, such as cell adhesion and migration, has gained increasing importance. A hypothesis has been raised on whether cells with more flexibility or deformability have a greater ability to migrate in three-dimensional confinements, including extracellular matrices[203]. These are the extensive network of proteins and other molecules that surround, support, and give structure to the cells and tissues within the body while influencing many cellular processes, including migration, wound healing and differentiation[204]. On the other hand, there is also a counterargument that challenges this theory, given the wide variations in the genetic and biochemical phenotypes of cancer cell types and the possibility of cancer cell type-specific variations, which may not allow for a general rule between greater deformability of cells and increased migration or invasion of confined 3D extracellular matrices to be formed reliably. Much more research is needed to improve this technique, but its potential use can be anticipated.

Endoscopes

We've discussed some methods of observing and manipulating microscopic, even nanoscale molecules like cells, but how about larger systems like organs? This is where endoscopes can come in. These and other optical fibre medical instruments comprise optical fibre bundles. Endoscopes are specifically used to see inside a human body, allowing physicians to see inside the body without requiring surgery. With an endoscope, we are given the capability to view lesions that an X-ray is unable to detect, which helps diagnose medical conditions[205]. Endoscopes use tubes with a thickness of merely a few millimetres to transport light in one direction while oppositely transferring high-resolution images, allowing for minimally invasive surgical operations[206].

The fibrescope is the main component of the endoscope. These are flexible optical fibre bundles with a lens and an eyepiece on opposite ends, which are used to investigate and see into small, hard-to-access areas, including the insides of locks and machines and the human body[206]. Using optical fibres allows for the light and the image to be directed flexibly around corners in the body, for example. A similar device is a borescope, which essentially performs the same function, but while a fibrescope is flexible, a borescope may be designed to be rigid. Flexible optical fibres are affixed to the probe head of fibrescope equipment, allowing operators to rotate the viewing lens head with greater flexibility. This broadens the viewing field so that the desired parts can be examined. Meanwhile, a rigid borescope would allow a straightforward or right angular view of the object, offering less flexibility[207]. Here, we focus on the fibrescope in endoscopy.

Endoscopy includes gastroscopes for the stomach, bronchoscopes for the lungs, and eye orthoscopes. Each scope may have different lengths or flexibilities depending on the body part used, but their basic function is the same. The majority of endoscopes are tiny, millimetre-thin tubes with an end that is equipped with a tiny camera, image sensors, and a strong light[208]. A natural opening introduces the scope, such as the rectum for a sigmoidoscopy or the mouth during a bronchoscopy[209]. Typically, an endoscope also has a channel through which the doctor can insert tools. These tools collect tissue or provide treatment. For example, flexible forceps are tong-like tools that take a tissue sample. Similarly, biopsy forceps remove a tissue sample or a suspicious growth[210]. Cell samples are taken using cytology brushes, and internal stitches are extracted using suture removal forceps.

Although there have been rudimentary endoscopic techniques for hundreds of years, the one that would serve as the model of our modern endoscopes was constructed in 1806 by German physician Philipp Bozzini[211]. He introduced a "Lichtleiter," or light conductor, for inspecting the human body's cavities and canals, although a mixture of professional rivalry and his untimely death stifled excitement for his discovery[212]. It was made of a tube that could be placed into a bodily cavity and had different attachments. The doctor could look within the cavity thanks to an inside candle and angled reflectors. Initially believed to be most helpful for laryngeal examinations, it later proved to be the forerunner of numerous urologic endoscopes[213]. Its design was later modified for gynaecological and urological uses[213].

This application can be seen with Antonin Jean Desormeaux, the French physician who developed the first successful open-tube endoscope[212]. He was the first surgeon to successfully operate using an endoscope, where the instrument was used to examine the urethra and the bladder[212]. A major improvement in endoscopy was enabled by the electric light bulb invented by Thomas Edison. Although these lights were used externally at the start, they were sufficiently capable of illumination to allow cystoscopy, hysteroscopy and sigmoidoscopy, examination of the nasal, and later thoracic cavities. For example, by 1865, Sir Francis Cruise regularly did these tests on human patients with his commercially accessible endoscope in the Mater Misericordiae Hospital in Dublin, Ireland. Later, smaller light bulbs became obtainable, enabling interior lighting.

Although Cruise also recorded the first thoracoscopic examination in a human, Hans Christian Jacobaeus is credited with initiating the first extensive documented series of endoscopic studies of the belly and thorax using a thoracoscope in 1910 and a laparoscope in 1912[212]. Then, in the 1930s, Heinz Kalk employed a laparoscope to diagnose gallbladder and liver disorders. In 1944, Raoul Palmer placed his patients in the Trendelenburg position following gaseous distention of the abdomen to perform gynecologic laparoscopy with reliability[212]. Laparoscopy is now a common procedure used in medicine. It is a surgical technique that looks at a woman's pelvic organs and the organs in her belly or abdomen[214]. As with other endoscopies, Laparoscopy uses a narrow, illuminated tube with a built-in video camera, also known as a laparoscope[214]. It is inserted into a minor abdominal incision. This provides an alternative, less invasive method compared to traditional open surgery, which comes with a larger incision and longer recovery, leading to more possible complications. However, laparoscopy may not be suitable for all cases.

One field undergoing evolutions to endoscopy is the area of neurotology and otology, pushing the limits of endoscope purposes from purely observational to operative[215]. These are related disciplines focusing on the ear. Patients with hearing loss, infection, vertigo, or the need for ear reconstruction are seen by an otologist[216]. Neurotology is a similar field but adds the knowledge and expertise to work within the skull on the specific portion of the nervous system and the brain related to balance and hearing[216]. Endoscopic ear surgery has become increasingly popular for adults and children over the last ten years to treat middle and lateral skull base diseases[215]. Another method for surgical procedures was the binocular microscope, which revolutionised otology by allowing illumination and magnification of the surgical field and freeing up a surgeon's hands to enable two-handed dissection. However, this came with its disadvantages. Due to the dimensions and shape of the external auditory canal (EAC) and the speculum, the term used for medical tools for investigating body orifices, the microscope provides a limited line-of-sight image with a narrow field of view. Therefore, building a parallel channel through the mastoid is frequently required to reach middle ear areas that are not visible under a microscope[215]. Postauricular techniques necessitate additional bony drilling and soft tissue retraction to surgically reach behind the external ear structure, resulting in either dysesthesia or anaesthesia in the auricle, keloids, hypertrophic scars or wound infection[215].

On the other hand, with endoscopy available, many of these limitations can be circumvented by offering a non-line-of-sight view that is less constrained by the size and form of the port, with illumination and image capturing occurring near the surgical field[215]. This is especially beneficial for patients with tortuous or small EACs. Additionally, the broad angle view and improved depth of field of the endoscopic picture enable surgeons to "look around corners" and into more hidden parts not visible with a microscope[215]. If the patient has a tiny ear canal, EES has specific benefits over line-of-sight microscopic ear surgery (MES)[215]. Overall, by using EAC as a surgical corridor with minimal access and eliminating the need for additional soft tissue retraction or bone drilling, EES can completely transform the fields of neurotology and otology[215].

Better visualisation and access are not the only benefits of the endoscope[215]. The cost-friendliness and improved ergonomics compared to MES also put EES at an advantage in addition to being a valuable tool in teaching since everyone shares an identical view of the surgery[215]. Extended periods of bad posture and poorly built operating room setups have been linked to an elevated risk of pain in the neck and back among surgeons, particularly those who perform prolonged microscopic work. In a UK survey of 325 ENT (ear, nose, and throat) physicians, 72% reported back or neck pain[215]. Of all ENT subspecialties, otologists reported the most pain, likely from constantly working with microscopes, which can demand hours of still hand, neck, and back positions.

While the operating room setup for EES is similar to regular MES, extra considerations must be taken to maximise body mechanics and facilitate "heads up" ear surgery[215]. To maintain a neutral, comfortable position for the neck, the mounted video screen or endoscopic video tower must be positioned facing the surgeon directly, as near as possible to eye level[215]. Instruments should be passed over the patient's torso by the scrub nurse and scrub table, which should be placed adjacent to the video monitor[215]. The anesthesiologist and surgeon should face each other for easier communication during the procedure. Research indicates that optimal placement of monitors and surgical settings enhances performance and efficacy, potentially improving both patient safety and cost-effectiveness[215].

That is not to say that endoscopes have no drawbacks of their own as well. Downsides include the requirement to complete all dissections one-handed with suction available on the other hand and the dependence on motion parallax for depth perception assessment[215]. Surgeons not educated in endoscopy face a steep learning curve and, in most cases, limited experience during their training[215]. Fortunately, as EES has grown in popularity, so have the training opportunities.

In otology, the popularity of otoendoscopy can be seen in its expansion in utility, advancing from observational to operative[215]. At first, endoscopes were used only in otology to depict ear anatomy. Later, they were used in MES to look for diseases that remained after a cholesteatoma resection[215]. This is the surgery to remove a cholesteatoma, an abnormal clump of skin cells located deep within your ear[217]. Though uncommon, they can potentially harm the sensitive inside structures of your ear that are vital for balance and hearing if left untreated. Following the introduction of EES in the late 1990s, the endoscope transitioned from an auxiliary observational instrument to the primary operating instrument[215].

The use of EES in treating middle ear disease in adults and children has increased dramatically during the past years. The treatment of cholesteatoma remains the most extensively researched use of EES, but non-cholesteatoma otologic endoscopic procedures have also seen a recent wave of research[215]. Applications of EES have increased dramatically thanks to these investigations. Perforations in the tympanic membrane and chronic ear conditions such as cholesteatomas and otosclerosis are notable examples of otologic indications[215]. Because of EES, surgeons can now do more trans-canal dissections for both simple and complicated middle ear conditions. Endoscopic-assisted transmastoid excision for chronic ear illness can help prevent a canal wall-down approach when necessary for a postauricular approach[218]. On each side of your skull behind your ears is the temporal bone with a bony protrusion known as the mastoid process[219]. It is relevant to ear issues because its location makes it susceptible to pain when experiencing problems such as a middle ear infection. Transmastoid excision is a less invasive surgery that keeps the ear canal intact with less damage to the mastoid. Children may find mastoid care particularly difficult, but a transcranial approach can aid in preventing recurrent postoperative debridement suture removal, mastoid dressing changes, and water precautions[215]. Ontologists have been very interested in ear canal surgery using an endoscope, called transcranial endoscopic ear surgery, for these reasons over the past ten years[215, 220].

Optical biopsy

Moving on to another technique. A biopsy is a common medical procedure to remove a cell or tissue sample from the body to be tested in a laboratory[221]. A biopsy is used for people experiencing certain signs and symptoms to determine the presence or extent of a disease. The word "biopsy" was first coined as a medical term by Ernest Besnier in 1879, but M. M. Rudnev performed the very first diagnostic biopsy in 1875 in Russia[222]. Until the years leading up to 1900, biopsies were simply used occasionally for histologic procedures utilising living organisms and tissues that are available for examination and observation[222]. During the 20th century, there began to be limited use of biopsy[222]. Still, because of its increasing acceptance, the approach is now largely used in almost all clinical disciplines, including oncology.

There is currently an alternate method for doing a so-called optical biopsy. Optical biopsy is simply a newer tool that allows for diagnosing diseases like cancer and atherosclerosis using spectroscopy at certain wavelengths without requiring the removal of bodily tissue[223]. "Optical biopsy" is technically incorrect because it means tissue is not actually removed, as "biopsy" refers explicitly to tissue removal[223]. Nonetheless, it is generally known to refer to the noninvasive, in situ, in vivo, real-time tissue diagnosis performed through some kind of optical measurement, frequently a kind of spectroscopy[223]. The treatment is often accompanied by endoscopy or fibre optics. Since endoscopes can provide a visual examination of the body, as previously indicated, these optical measurements may allow for "guided biopsy," decreasing the number of tissue samples taken while increasing the likelihood of sampling a diseased site[223]. Reducing the necessity for surgical biopsy tissue sample removal is one of the primary driving forces behind this research. Furthermore, the patient's emotional trauma while waiting for a response may be lessened by the immediacy of the diagnostic information.

The advantages of optical tissue diagnosis can be substantial for gastrointestinal (GI) tract illnesses, for instance. A higher risk of developing cancer has been linked to several GI tract conditions, such as colitis, colon polyps, and Barrett's oesophagus[223]. These conditions are usually monitored with tissue biopsies and endoscopic exams performed at least once a year. Up to 20–30 "random" biopsies could be performed in a single session[223]. This expensive and time-consuming operation involves some risk to the patient. Before the biopsy tool may be reinserted for the subsequent conventional biopsy, the specimen must be removed from the endoscope and the biopsy tool retracted[223].

Conversely, by merely shifting the probe tip's location, an optical diagnostic probe might be moved successively between sites, recording each measurement in a split second[223]. Surgical biopsies can be carried out at the site of illness identification. Consequently, optical biopsy minimises tissue damage and saves time[223].

Photodynamic therapy (PDT) and photothermal therapy (PTT)

Proceeding to focus more on treatment rather than detection, photodynamic therapy (PDT) is an example of light being used in treating disease. Photodynamic therapy is where a medication known as a photosensitiser is combined with light energy in a two-stage procedure to kill malignant and precancerous cells upon stimulation by light[224]. Agents known as photosensitisers absorb certain wavelengths of light, most likely from a laser and transform them into usable energy. In PDT, cytotoxic chemicals are produced when photosensitisers, light, and molecular oxygen combine, resulting in cell death. There are numerous photosensitiser medications on the market now to treat a range of illnesses[224], ranging from acne to psoriasis to age-related macular degeneration to several forms of cancers, including those of the oesophagus, bladder, pancreatic, brain, lungs, and skin. In addition, PDT can also help with the treatment of fungal, viral, and bacterial infections. Studies have indicated that this light-based treatment can activate the immune system, providing your body with another tool to help destroy precancerous and cancerous cells[224]. More people are realising the benefits of this approach as a treatment for some localised cancers, meaning those that haven't spread far from their original site.

The US Food and Drug Administration (FDA) approved several photosensitising medicines to treat specific cancers and precancers[225]. First, the two most widely used are porfimer sodium, also known as Photofrin, a commonly used and researched photosensitiser triggered by a laser red light. It is being studied for various cancer types and has been accepted by the FDA for the treatment of specific forms of lung and esophageal cancers[225]. Secondly, aminolevulinic acid, or ALA or Levulan is a topical medication. It can treat actinic keratosis, a skin condition that can progress into cancer, used exclusively on the scalp or face[225]. Instead of laser light, this medication is triggered by a specific blue light.

On the topic of how the drug is administered, Depending on which region of the body needs to be treated, the photosensitising medication can be applied topically or injected into a vein. Over time, the cancer cells absorb the drug. The region has to be subjected to illumination. The medication reacts with the light to produce a particular type of oxygen molecule that destroys the cells[225]. PDT may prove useful when you cut off the blood arteries that provide nutrients for the cancer cells and trigger the immune system to fight the cancer[225]. The time that passes between applying light and administering medication is referred to as the "drug-to-light interval." Depending on the medication taken, it could take a few hours or several days. Certain types of lasers or light-emitting diodes provide the light needed in PDT[225]. The cancer type and location throughout the body determine the type of light that is employed. Also, PDT is often performed as an outpatient operation, meaning no hospital stay is required[225]. However, it can occasionally be combined with chemotherapy, radiation therapy, or other anti-cancer medication[225].

Research has demonstrated that PDT is an effective treatment for certain types of cancer and precancerous conditions, just as radiation therapy or surgery, as it offers several benefits. For instance, when administered correctly, it can be precisely targeted and usually has no long-term negative effects. Compared to surgery, it is less intrusive, often only requires a brief time, and is performed as an outpatient procedure[225]. In contrast to radiation, PDT can be performed repeatedly at the same location if necessary, and following the site's healing, there is typically little to no scarring[225]. It also frequently costs less than alternative cancer therapies.

However, PDT has its limitations as well. PDT's primary drawback is its limited ability to treat only regions where light can be accessed[225]. This means that issues on or just beneath the skin, as well as in the organ linings accessible to light sources, are the main conditions it treats. PDT is also ineffective for big cancers or malignancies that have penetrated deep into the skin or other organs since light cannot pass through bodily tissues very far[225]. In other words, cancers that have spread to multiple locations cannot be treated with PDT. Although there are typically no major side effects, different people may experience various short-term adverse effects[225]. PDT, for example, is not advised for those with specific blood disorders. One major side effect of PDT medications is that they make users extremely sensitive to sunshine or strong light for a while[225]. As such, precautions need to be taken after administering the medication. On the area of the skin where the medication is administered, these PDT light-induced reactions may become apparent. They typically cause burning or tingling sensations along with redness. Patients will need extra care to avoid exposing the regions of their scalp and face that were treated to light for a while following the procedure[225]. It is recommended that patients receiving these therapies stay indoors, avoid bright light, and wear protective clothes while they are outside[225].

A very similar procedure with a similar name is PTT, photothermal therapy. This is also a common and more recent innovation in phototherapy for cancer treatment. Although not necessary for PTT, they both use light and either endogenous or exogenous absorbers, meaning the photosensitisers are applied topically or injected/ingested, respectively[226]. However, their mechanism of killing the tumours is different. While PDT generates cytotoxic reactive oxygen species, PTT converts the light into heat energy, increasing the local temperature to kill the cancer cells by causing cell membrane damage and protein denaturation to induce rapid cell death[226]. PTT can also be used as an imaging technique to track the inflation of PTT agents in tumours in real-time or temperature monitoring and visualise the precise position of cancer cells[226]. This can be accomplished with MRI. For instance, recent advancements in PTT have used tiny optical fibres to deliver NIR light into solid tumours. Using MRI or ultrasound guidance, the treatment selectivity and safety are improved by ensuring the exact placement of the fibres[226]. However, due to the cost requirements of MR thermometry, other optical approaches, including diffuse optical tomography and photoacoustic imaging, both previously discussed, have shown promise as possibly more affordable ways to track the coagulation front with greater sensitivity and specificity[226].

Several key factors must be considered to provide a more comprehensive comparison between PDT and PTT. First, due to photosensitizers buildup and targeted light delivery, PDT offers inherent selectivity advantages. This makes PDT effects more selective compared to PTT. Furthermore, the neighbouring nerves and collagen structures can be preserved because PDT is a non-thermal method[226]. In contrast, PTT produces a temperature gradient because its selectivity relies mainly on localised light delivery. Insufficient heat containment may lead to thermal damage outside the designated photocoagulation zone, raising the possibility of risky outcomes. PDT also offers the advantage of targeting infiltrative and unresectable tumour components, which are areas that cannot be surgically removed[226]. This is facilitated by the short diffusion distance of reactive oxygen species in PDT. However, it's important to mention that in some cases, the accumulation of photosensitisers in certain body parts, such as the eye or skin, may cause photosensitivity, as explained earlier, which can compromise patient safety and add extra complications to the treatment logistics[226].

Furthermore, distinct types of light sources are each required by PTT and PDT. Fluence rate refers to the radiation incident per unit surface area per unit time. PDT can utilise a range of light-emitting diodes, comparatively cheap low-power lasers, and even natural daylight because of its low total fluence and fluence rates[226]. In contrast, PTT necessitates more sophisticated fibre cooling, online temperature monitoring, and more costly, higher-energy lasers. While it may seem that PPT equipment has higher upfront costs, PDT can incur expenditures associated with the photosensitisers, the requirement for separate appointments for its administration, and lengthier operating room durations for PDT irradiation[226]. Interestingly, due to their difference in mechanism of action, there has been evidence to suggest that in combination, possibly with immunotherapy as well, the processes may synergise to potentially improve the efficacy of the treatment, although more research is needed.

There is another specific treatment route in development called photoimmunotherapy for cancer treatment. This uses antibody–photo absorber conjugate (a combination of antibody and photo absorber) that attaches to cancer cells. Antibody-drug conjugates are substances made up of monoclonal antibodies, man-made proteins that act like human antibodies in the immune system, chemically linked to a drug. The monoclonal antibody binds itself to certain proteins or receptors of particular cell types[227]. When the linked medication enters these cells, it destroys them without endangering healthy cells[227]. Likewise, antibody-photo absorber conjugates can target cancer cells so that the photo absorber binds with those cells. Then, in near-infrared photoimmunotherapy (NIR-PIT), when exposed to near-infrared light, the cancer cell is killed as the cells swell and eventually rupture. This NIR-PIT approach is the fastest-growing PIT approach. During this procedure, not only does the cell swell, causing the membrane to rupture, also known as immunogenic cell death[228], but it also results in triggering vigorous anti-tumour immune responses. This is primarily attributed to releasing damage-associated molecular patterns and tumour-associated antigens following PIT[228], which activate the body's immune response against the tumour. However, the effectiveness of this immune response can be suppressed by the immunosuppressive nature of the tumour microenvironment, particularly in cases of metastatic tumours. To enhance the antitumor immune response, a combination of PIT with immune checkpoint inhibitors and immunoadjuvants, other agents used in antigen conjugation, can be employed as a synergistic treatment against tumour cells. Notably, NIR-PIT offers advantages over traditional PDT, including independence from ROS generation, higher efficacy, and minimal to no side effects on surrounding healthy cells[229]. These characteristics make NIR-PIT a highly promising approach for the selective and localised destruction of cancer cells.

Laser ablation

A more straightforward treatment method would be laser ablation. Ablation means removing or eliminating tissue or a body component, so laser ablation, also referred to as photoablation, is the technique of using a laser beam to remove material from a solid surface. When the laser flux is low, the material is heated by the absorbed energy from the laser, causing it to either evaporate or sublimate[230]. At a higher laser intensity, the substance is normally transformed into plasma[230]. In medicine, this can be implemented as a treatment to remove tumours and other lesions. Using an MRI-guided laser probe, it delivers light and thermal energy to destroy unwanted cells. Besides destroying or ablating tumours, it can also reduce the number and severity of seizures or cure epilepsy completely, as well as improve symptoms associated with treatment-resistant obsessive-compulsive disorder for those whom medication doesn't seem to be as effective[231].

In some clinics, an advanced surgical instrument known as ROSA Brain assists the surgeon in precisely placing a probe to reach the brain lesion. ROSA is a robotic technology arm that helps surgeons plan and carry out intricate neurosurgery treatments in a minimally invasive manner. Although ROSA can be useful in many other neurosurgery operations, its main application is treating epilepsy[232]. The light-emitting probe is a thin, flexible tube. A minor incision that is a little wider than the probe is made in the scalp and skull by the surgeon, who then inserts the probe[233]. After that, the patient is placed into the MRI machine. The surgeon verifies the precise location of the probe tip in the brain using the MRI monitor. The location and temperature of the heated tissue are displayed on a computer monitor[233]. This aids the surgeon in determining when to cease treatment and how much to administer[233]. The actual laser treatment requires only a few minutes to complete. It's the careful setup before treatment that requires a longer time; around 4 hours are spent under anaesthesia[233].

Here are some of the key benefits of laser ablation. We can compare it to another option for such epilepsy treatments, an open brain craniotomy, which encompasses temporarily removing a portion of the skull to allow for surgical access to the brain[234]. Laser ablation has been shown to be a safer and less invasive process than that. This is because there is less possibility of harming healthy tissues during the procedure and the laser is inserted through tiny bone incisions and holes[234]. It requires less time for rehabilitation, surgery, and hospital stays[234]. Other significant benefits include the capacity to treat places in the brain that are too deep for a surgeon to access, the elimination of the requirement for patients to shave the surgical location so they can keep their hair, reduced pain and scarring following the treatment, and overall fewer complications than with standard craniotomies[234]. The likelihood of damage to the patient's vision, mobility, memory, language, and other brain functions is reduced by the more precise treatment of the lesion[233]. Especially for children, it takes less of a toll on their bodies, leading to faster recovery and less pain[233].

Nevertheless, as with any operation, laser ablation carries some dangers, such as the potential for infection or damage to a healthy brain region[233]. Although they are uncommon, some problems that have been recorded include nerve injury, brain haemorrhage, and eyesight constriction[234]. For now, these are inevitable since side effects are generally a given regardless of how successful a treatment is.

Low-level light laser therapy (LLLT)

Photobiomodulation therapy, also referred to as low-level light laser therapy (LLLT)[235], is another fascinating technology created as part of the treatment procedure. LLLT is not a thermal therapy, unlike many other therapies. Rather, it alters photochemical reactions in cells, known as photo bioactivation. Some understanding of biological mechanisms is required to comprehend how LLLT functions. Firstly, the energy that our cells' mitochondria make is known as ATP or adenosine triphosphate. This is made with oxygen and glucose. The electron transport chain (ETC) and Kreb's Cycle are two of the processes that occur inside the mitochondria, the cell's power plants. Cytochrome C Oxidase (CcO) is an enzyme found near the end of this ETC.

Due to factors like injuries, excessive sugar in our diets, and cigarette smoking, cells can experience stress. This will result in a rise in the formation of free radicals, which are usually unstable and extremely reactive molecules. In particular, a molecule known as nitric oxide (NO) connects to the enzyme CcO first rather than oxygen since it has a greater affinity[235]. As a result, there are fewer or no places for oxygen to attach, meaning cells cannot create as much ATP as they need to operate properly, which leads to even more stress and starts producing reactive oxygen species (ROS), which are extremely volatile free radicals[235]. Normally, the cell uses antioxidants and rest to control and neutralise ROS. Oxidative stress arises when this process is not completed. Another scenario of oxidation would be a metal corroding. If left unchecked, it can eventually cause holes in the metal, and similarly, oxidative stress is recognised as the underlying cause of most diseases and degenerative problems in our cells. It also contributes to the ageing process and the inflammatory stage of both acute and chronic injuries.

This is the problem that LLLT is used to address. When the proper wavelength and intensity of light are applied to the stress area, NO is quickly displaced from the receptor site CcO, creating local vasodilation. This is how LLLT generates its physiological and therapeutic effects. Thanks to the NO being removed from the site, oxygen is able to bind and start generating ATP once again. This aids in reducing oxidative stress, higher ATP, and enhanced circulation, promoting healing, pain alleviation, and managing the inflammatory response[236]. These effects depend on the redox state and dosage. It has been repeatedly demonstrated that after photobiomodulation therapy, nitric oxide is released, ATP is raised, and oxidative stress is decreased in hypoxic or otherwise stressed cells.

Overall, we have listed and explained a diverse range of state-of-the-art solutions developed to advance our capabilities in detection, imaging, diagnosis, treatment, research, and practically any area within medicine. To sum up, the general advantages of employing photonics or optics-based technology include reduced invasiveness, higher accuracy or specificity, and expanded in-vivo abilities such as imaging or measuring the properties of structures as small as cells or even further. Perhaps most notably, they have aided the whole oncology procedure by studying how light affects tumours. Although they are all mostly rooted in the fact that different molecules interact with light and photons differently, as you can see, the possible variations and applications are endless. Certain properties can still be improved to help these technologies reach their full potential, such as sensibility and penetrability, which are sometimes limited due to light scattering.

Artificial intelligence has also been incorporated into biophotonics. For example, machine learning and AI systems can be used to carry out image processing and analysis, yielding faster and more accurate results than if this process were repeated manually for several images. An incredibly useful application for this would be in diagnosis. A computer can be trained through machine learning to recognise characteristics or indicators of diseases, identify their presence in the image, and efficiently diagnose the patient. Speed is a crucial benefit as with traditional histopathology, the tissue and generated analysis reports may take several days to process. This is particularly evident in critical time-sensitive situations, such as the intraoperative evaluation of breast tissue during surgical oncology procedures. Eventually, an automated diagnosis system could be developed to reduce delays in treatment due to diagnosis time. This couples with the advantage of in-vivo imaging from photonics devices, where real-time imaging is achieved rather than having to perform biopsies or other procedures for internal structures.

With the development of our current technologies, a drastic shift has occurred in biomedicine practices. Over the next decades, there will most likely be as big an improvement once again where new properties and techniques of light can be taken advantage of.

Chapter 5: Other prominent fields

Although biophotonics may seem to focus on medical applications, it includes various applications from food safety to environmental conservation. Essentially, any photonics technology connected to biology can be classified as part of biophotonics. Each area mostly uses the same fundamental technologies explained in the previous chapter, such as fluorescent spectroscopy, optical fibre sensors and laser-based techniques. This chapter will discuss how these same techniques are adapted and used in different areas and introduce some new photonics methods when they exist.

We shall start with agriculture. Agriculture has always been an extremely important industry in our world, with the shift from hunting to cultivating crops around 12,000 years ago transforming our food industry. As of 2023, agricultural exports from the US were worth $174.9 billion, with farming contributing approximately 1% of the country's GDP[237]. However, with the increasing global population, climate change, and diminishing resources, farmers face challenges requiring more precise and sustainable farming methods than traditional methods. These traditional practices can be imprecise, leading to overuse of resources and poorer crop yields. Another factor is climate change, which leads to unpredictable weather patterns, such as sudden variations in temperature and precipitation. It could also increase the spread of diseases and pests. All this, in turn, leads to a higher susceptibility to crop failure. As for the growing population, the UN predicts that the global population will reach 9.6 billion by 2050, from around 8.1 in 2024. To compensate for this, an increase of around 70% in food production by 2050 will be imperative to keep everyone decently fed[238].

Biophotonics offers solutions to these problems. It is mostly used in cooperation with machine learning to reach its full potential. Since biophotonics is the study of interactions of light and matter found in biological systems[238], it has potential applications for purposes such as crop monitoring and disease detection, allowing for more precise crop targeting. Machine learning is then leveraged to analyse the data collected by the photonics devices and make predictions on the weather or crop yields with greater precision, which can help optimise crop yields and accelerate the processes of farming operations. Combining the capabilities of each technology, farming will experience a rise in productivity and reliability as it will have concrete evidence and data rather than be dependent on guesswork.

This approach to management that places a strong focus on monitoring, measuring, and responding to changes in fields, crops, and animals in almost real time is known as precision farming[239]. These techniques can lower expenses, especially labour costs, by reducing manual checks or procedures, increasing the process input efficiency, and, most importantly, raising crop yields and animal performance[239]. Photonics and machine learning act as crucial building blocks in most of the technology needed to enable precision farming. Bioengineering, robotics and automation, imaging and sensor technology, digitalisation and big data analytics are examples of the technologies that can provide solutions for precision farming[239]. All these would not be conceivable without photonics devices to detect, monitor, understand, and control them. Actually, various solutions already exist, but precision agriculture is still a relatively new concept with its emergence in the US in the 1980s. Numerous innovative solutions have been developed, and precision agriculture has clear potential for growth. As of 2023, the photonics for precision agriculture market was estimated to be worth 4.6 billion euros globally, and it was projected to nearly double to 9.1 billion euros by 2027[239]. This translates to around 15% yearly growth. Requirements for these technologies to be successful involve cooperation between farmers and physicists or engineers to make use of each of their expertise. Secondly, the diversity of crops and animals is incredible. For instance, there are over 800 farmed races of cows and over 10,000 types of apples[239]. Every crop variety and animal race has its own unique physical traits, diseases, and pests, and they are all adapted to particular climates and types of land[239]. As a result, the problem is developing versatile and adaptable photonic sensors or equipment that achieve the same efficiency no matter the variety used by simply adjusting the operating parameters using software[239]. If this ability is obtained, the same hardware can be used across different species at a reasonable cost and on a large enough scale[239].

In traditional agriculture, farmers would manually fertilise, sow, or apply pesticides. While their experience allows them to do so with high precision and efficiency, human limits mean that it is not feasible to continue with these traditional methods to feed 10 billion people in the future. Even now, the pace of agricultural expansion is not enough to sustain the world's population. In 2021, 11 percent of the world's population was undernourished, and 3 billion people could not afford a healthy diet in 2019[239]. Needless to say, agriculture will not be able to continue to expand at the current rate to meet the nutritional needs of every individual of the highest possible quality.

To cope with this, agriculture must adapt and automate its systems. For instance, the expanded croplands can be ploughed, harrowed, seeded, and fertilised by machines with high working widths and fast speeds. Cleaning stables, feeding, or milking animals can also be automated through steering devices and robots. This is where photonics and its solutions come into play. Photonics' sensor and imaging technologies can be mounted on drones, tractors, or machines to enable them to move autonomously, explore their environment, make decisions, and perform their tasks. These sensors remove the need for farmers to constantly manually monitor their fields and animals. In other words, the process of evaluating crop quality, sorting crops by quality standards and removing unwanted ones will all be automated. For image processing, as with the methods for remote detection developed for medical applications, photonics is vital to perform advanced analysis of diseases, soil and water conditions, crop and fruit quality, and many more without contact[239]. Data analysis is powered by machine learning algorithms to improve the forecasting of the climate and soil, equipment performance optimisation, and remote control in field monitoring. With the high volume of data generated from precision farming, communications involved in data centres depend on the use of fibre optics and transceivers, which are the components that convert light into electricity and vice versa[239]. This allows the data to reach the computers and be processed. Using fibre optics also improves data sharing to alert other agriculture stakeholders, including neighbouring farmers, banks, seed suppliers, etc.

As for bioengineering, which refers to the branch of engineering focusing on biomedical technology and biological systems. This ranges from the engineering of actual machines like MRI scanners to artificial parts like artificial joints to genetic engineering or biochemical engineering. Regarding genetics, the research relies heavily on microscopy techniques, including the more advanced models described in the earlier chapter. Genetics would not be possible without PCR analysis read by photonic technologies[239]. PCR, the polymerase chain reaction, is a commonly used technique that rapidly produces millions or even billions of duplicates of a particular DNA sample[240]. This enables scientists to augment a tiny sample of DNA to the point that it can be thoroughly studied[240]. The significance of this discovery can be seen in the 1993 Nobel Prize in Chemistry being awarded to Kary Mullis, the scientist who invented PCR in 1983. Evidently, PCR is important. Nowadays, it plays a crucial role in numerous processes involved in genetic testing and research, such as the identification of infectious organisms and the study of ancient DNA samples. Also, the ability to assess observable or measurable attributes of an organism, also known as phenotyping, is based on advanced imaging, which again requires photonics. With the help of these techniques, bioengineering allows seeds and chemicals to be chosen according to external conditions, and seeds can be modified to increase resistance to particular climatic or agricultural circumstances.

3D laser scanners

Beginning with lasers once again, 3D laser scanners are used to measure data about plants. Specifically, terrestrial laser scanning, which uses active remote-sensing laser rangefinders, can serve as a tool to monitor plant growth, which can then indicate crop performance with 3D lasers. There are fixed scanners that can be situated on a tripod to rotate with a computer connected to store data. In comparison, mobile terrestrial laser scanners are usually made airborne by mounting them on UAVs, commonly called drones. However, they are both versions of LiDAR technology, so their fundamentals are identical. They work by emitting millions of laser pulses. The dimensions and spatial relationships of the plants can be accurately measured and calculated by measuring how long is needed for the reflection of the laser pulses to travel back to the receiver. With this time and knowing the speed of light, the object's location can be calculated. Newer generations of this technology have been developing to yield each point's passive colour, containing the reflected pulse's red, green, and blue values.

With this process, a 360-degree scan is taken, and the millions of data points captured from the surfaces of the objects result in a precise point cloud of the scanned area. A point cloud is essentially just a collection of 3D coordinate points, meaning each piece of data contains XYZ coordinates[241]. They can be used to represent 3D objects for modelling, and they are among the sources used in geographic information systems to create a digital elevation model of the landscape. Most software will allow different surveys to be merged into a single point cloud. This way, the areas hidden from one survey can be complemented by scans where the lasers measured those "shadow" areas. This gives a very high point density to allow for greater resolution so the surveyed scene can be modelled with high accuracy and detail in all three dimensions. In ideal conditions, a distance of up to 200 metres can be measured, although 50-100 metres is more common. With the latest technologies, a full 360-degree scan can be carried out in less than 4 minutes.

For agricultural purposes, using scanners for 3D reconstruction allows for automatically measuring plant growth, non-destructively and non-invasively. A simple approach would be to take images with a commercial digital camera to observe plant height[242]. In more complex approaches, stereo cameras offer a way to reconstruct 3D images by analysing multiple images from different angles or taking depth images[242]. Stereo cameras are different from conventional cameras in that they use two digital cameras together to capture the same scene from two viewpoints. This allows the camera to replicate the stereoscopic vision of human eyes. This refers to the fact that our eyes view objects from two slightly different perspectives, and combined, this creates an impression of depth and solidity.

An additional specialised technology known as SLAM is required for mobile terrestrial laser scanning. SLAM stands for simultaneous localisation and mapping and is a broad term across various technological processes and algorithms that allow robots to travel through unfamiliar areas autonomously without requiring a map[243]. The issue arises because autonomous navigation necessitates both the machine's location in the environment and the creation of a map of that environment. This is challenging to accomplish because the machine can't estimate its own location without a map of the surroundings[243]. However, it needs to know its location to create the map. This leads to an endless circle of dependencies. This is the complication SLAM aims to solve. As stated, there are many approaches to SLAM, but some generalisations can be made to demonstrate the basic concept behind the technology. In simplified terms, when the robot begins to operate, SLAM technology utilises data from the robot's onboard sensors[243]. It runs them through algorithms for computer vision to identify elements in its immediate surroundings. This allows SLAM to create a basic map and estimate the robot's initial position. As the robot moves, SLAM uses the initial estimate, gathers new sensor data, and generates an updated and more accurate position estimate[243]. With this improved updated position estimate, the map is concurrently updated. By continually performing these actions, SLAM tracks the robot's trajectory and creates a detailed map of the area[243]. Unfortunately, though, aerial scanning tends to be less accurate compared to fixed scanners, so even more research is needed to improve precision and continue to reduce costs so more farmers can adopt these techniques.

Hyperspectral imaging (HSI)

A similar technique is hyperspectral imaging (HSI). This is the combination of spectroscopy and digital imaging. This technique involves gathering and analysing data across different parts of the electromagnetic spectrum, including those outside the visible range, to capture the spectrum corresponding to each pixel in an image. Every substance and compound has a unique spectral signature because of how they respond to light, so each pixel in the resulting hyperspectral image represents a distinct spectrum. By examining the distinct spectral signatures, we can identify objects and materials in the image[244]. In HSI, spectral information is incorporated into every image pixel, creating a three-dimensional data cube by extending the two-dimensional spatial representation with an additional third dimension of values[244]. This data cube, also known as a hypercube or image cube, encompasses data from the reflectance, absorption, or fluorescence spectrum for every pixel in the image. With HSI, measurements between thousands to hundreds of thousands of spectra are combined to build an enormous hyperspectral data cube[244]. HSI differs from multispectral imaging, which is a collection of layers of images of the same scene and different particular wavelength bands rather than continuous wavelengths like in HSI. The standard definition states that to be counted as HSI, the camera must have more than 100 bands, while multispectral imaging only makes use of much fewer bands[245]. Hyperspectral imaging involves collecting and analysing data from many different narrow and connecting bands throughout the electromagnetic spectrum. The reason why hyperspectral cameras can produce smooth spectra is because this process yields high-resolution spectra for every pixel in the image[245].

In contrast, multispectral cameras generate spectra that resemble staircases or saw teeth, lacking the ability to capture fine spectral details[245]. In circumstances where the user does not require a wide range of spectral bands, multispectral imaging will suffice. Still, in situations like precision farming, where accuracy is required, more bands will need to be measured.

The acquisition of spectral information in hyperspectral imaging relies on an imaging spectrometer, also called a hyperspectral camera. Unlike a traditional RGB (red, green, and blue) camera that captures images using three bands of visible light, hyperspectral imagery enables examining object interactions across numerous bands, extending from wavelengths of 250 nm to 15,000 nm, including the thermal infrared range[244]. By separating the light of a scene into its distinct bands or wavelengths in the spectrum, the hyperspectral camera captures the spectral characteristics of every pixel in the picture[244]. This is how the camera can record each pixel's spectral details while creating a two-dimensional image of the scene.

Hyperspectral imaging brings together the advantages of digital imaging and spectroscopy[244], enabling the acquisition of both spectral and spatial data regarding an object's physical and chemical characteristics. This comprehensive approach offers more detailed insights compared to traditional spectroscopy methods. The spectral information obtained allows for classifying and identifying materials, while the spatial data provides information about the distribution and separation between materials within the object or scene[244]. In other words, hyperspectral imaging answers the questions of what the object is, where it is located, and when it was recorded[244]. An additional benefit of HSI is that because it includes wavelengths outside the visible range, identifying objects with similar physical or visual traits that humans can not differentiate is possible[244].

Relating to agriculture, to maximise crop management techniques and raise agricultural yields, HSI is capable of evaluating the health and productivity of crops by tracking soil moisture and nutrient content. Typically, agricultural surveillance is carried out with hyperspectral cameras installed on drones. It is used similarly to laser scanners. It allows for monitoring crop health as even minor alterations in the vegetation due to disease, stress, water scarcity, or nutrient deficiencies can affect the plant reflectance spectrum[246]. Environmental stresses include things like pollution and climate change. This is useful as some of these subtle changes are not immediately manifested physically, so farmers may miss invisible problems if they merely monitor manually[246]. With this information, farmers can intervene earlier and have better-targeted treatments to lower crop losses while increasing production.

Along with regular monitoring, they can gather a range of spectral information to help them make accurate schedules for planting, watering, and harvesting[246]. The efficacy of the fertiliser can also be analysed in terms of both type and quantity[246]. All this optimises yield, minimises resource waste, and boosts efficiency. By using the reflected spectra, the physical traits of plants can be analysed. Plant phenotyping analyses factors that impact the production of crops, such as plant growth patterns and leaf area. This valuable information assists breeding programs when creating enhanced crop varieties. Similarly, hyperspectral imaging differentiates crops and weeds according to their unique spectral characteristics. This analysis helps to optimise resource allocation, minimise overuse of herbicides, and develop tailored weed management tactics[246].

One main difference between LiDAR and hyperspectral imaging is that they are active and passive, respectively. LiDAR is active since it sends out a laser signal whose reflections are to be measured, while HSI and multispectral imaging simply measure naturally existing spectral signals from objects. These two techniques can be integrated to produce a monitoring system with expanded functionality.

While HSI can provide optical images with high temporal resolution with information, especially for differentiating different materials, other remote sensing technologies like LiDAR can complement these by enhancing or adding to these data with information on different formations, shapes, and geometrical features such as plant height[247]. By combining LIDAR and hyperspectral data, UAVs can possess synergised functionality whereby LiDAR data can be used to estimate terrain and vegetation vertically[247]. In contrast, hyperspectral data can help derive vegetation, exposed ground cover, and horizontal composition[247]. With this combination, the three-dimensional model and the map of surface properties produced are superior to those produced if they were used separately[247].

Spectrophotometry

One part of agriculture is aquaculture, which is farming aquatic plants and creatures like fish. According to the National Oceanic and Atmospheric Administration Fisheries, aquaculture is a significant part of agriculture, accounting for 50% of global seafood production for human consumption. Monitoring the water's conditions, including temperature, pH, and pathogen levels, is extremely important. Using human vision, one basic method of analysing water quality is adding a chemical reactant to a sample to compare its resulting colour to a reference colour sample[248]. Evidently, this method is prone to inaccuracies and offers only qualitative data, lacking the capability to provide quantitative information for in-depth analysis. Moreover, introducing chemical reactants contaminates the water sample, rendering it unsuitable for subsequent testing and analysis.

In contrast, spectrophotometry offers a non-destructive alternative to analyse potential contaminants in our water sources. Spectrophotometers can identify contaminants within a sample of water and measure its APHA colour, alternatively known as the Hazen scale or Platinum Cobalt scale[249]. Originally developed by the American Public Health Association to assess the colour of wastewater, the APHA colour scale has since been adopted for various industrial applications. It serves as a standardised test method for evaluating the colour of clear to yellowish liquids, so this scale is occasionally known as a "yellowness index"[249]. By measuring particular wavelengths of light, the spectrophotometer accurately determines the colour value, enabling the precise classification of chemical components and characteristics within the water sample[248].

A spectrophotometer is not the same as a spectrometer. Spectrophotometers are specifically designed for spectrometry, which involves the examination of colour spectra emitted by radiated matter[250]. In contrast, spectroscopy is a more general field that investigates the interactions of radiated matter[250]. While spectrometers detect and examine light waves to measure the physical characteristics of a substance across a spectrum, spectrophotometry requires specialised instruments capable of both spectrometer functions and observing colour changes[250]. This is why spectrophotometers incorporate a spectrometer, allowing them to fulfil all requirements for sample analysis. Combining these techniques makes it possible to study the interactions between particles, the makeup of the substance, and colourimetry.

A spectrophotometer water quality sensor measures light absorption in a liquid sample. This sensor comprises a detector, a sample compartment, and a light source. When light from the source enters the sample container, it reacts with the liquid's molecules so that while the molecules absorb some of the light, the remaining light is transmitted to the detector through the sample[251]. The detector quantifies how much light can pass through the sample and transmits this data to a computer[251]. Using specific wavelengths, the computer then uses the intensity of light absorbed to calculate the amounts of different compounds in the liquid. By using spectrophotometer sensors, it is possible to measure the levels of various components in water to ensure they stay within safety requirements. One such component is chlorine, commonly used in water treatment facilities as a disinfectant. Secondly, drinking water with high amounts of nitrate can be detrimental to animal health. Thirdly, increased phosphate levels in water bodies like ponds can trigger excessive algae growth, leading to decreased oxygen levels and damage to aquatic life.

Additionally, turbidity is how cloudy the water is and may be an indicator of suspended solids and particulates, which can negatively affect those consuming it. Hence, it is vital to measure and monitor turbidity levels as well. Some advantages of spectrophotometers include ease of use, high accuracy, reduced risk of human error, and durability. It can measure and identify water quality changes efficiently and continuously. To combat the issue of chemical reactants, this method is free of reagents and does not require the sample to be pretreated.

Fibre optic chemical sensors

Another main property that needs to be measured is the dissolved oxygen levels in water since this is how aquatic plants and animals respire. Fibre optic chemical sensors, which are used in food, agriculture, marine sciences, and biotechnology, allow for more compact and customizable structure designs and non-intrusive measurements. These chemical sensors may work by trapping an oxygen-sensitive fluorophore in a jelly-like material known as sol-gel, which is then applied to an adhesive membrane[252]. There is an indicator material which changes optical characteristics depending on the specific analytes while the electronic components measure the response. These can be independent oxygen monitoring systems or integrated into other sensing or electronic systems. Afterwards, machine learning algorithms can help with the processing of spectral data, especially if the client does not have the experience or desire to interpret this complex data. The installations within or above the tanks are linked to these intelligent data processors, so the spectral or optical data collected can be processed and transported, either in a miniature photonic chip or an optical waveguide. The combination can enable precision aquaculture, a subset of precision agriculture, and expand the usefulness of optical technologies used in aquaculture. Again, these advancements will improve yield, facilitate efficient and effective production, and maintain sustainability. The possible causes of harmful events on aquatic life can be identified more quickly, allowing critical information and breakthroughs to happen in aquaculture, which then increases the quality of food for human consumption.

Ultraviolet disinfection

The above is linked to the detection of pollutants in water. One method for the treatment of water is ultraviolet disinfection. More specifically, UV-C flashing is used. UV-C means ultraviolet-C light, also known as germicidal UV since it can kill or disable bacteria. There are three subbands within the UV spectrum. The difference is in their wavelengths. The longest wavelengths belong to UV-A radiation, whereas the smallest wavelengths are classified as UV-C radiation with UV-B in the middle. The short wavelength of UV-C is the reason for its hazardous nature. This is because photons with shorter wavelengths or higher frequencies carry more energy, which can more easily penetrate the skin into the cells to cause damage. Earth's ozone layer protects us from these harmful wavelengths by absorbing them while letting UV-A and some UV-B rays transmit through the atmosphere. Thus, the majority of UV rays that you encounter are UV-A, with very little UV-B. Actually, this technique works for more than just water; it extends to plants and even the air. With the shorter wavelengths, UV-B and especially UV-C can kill pathogens and pests without harming plants or animals. In 2012, the first commercial UV-C LED-based water disinfection system was introduced, although UV light has a long, well-established history of germicidal effects[253]. Not only is the saving in energy enabled by the use of LEDs appealing to many industries, such as water purification, but the extremely small sizes of these LEDs make them flexible so that they can create portable disinfection systems, for example. While UV-C (200-280 nm) has germicidal effects, UV-B (280-320 nm) is also shown to have antimicrobial effects, including dealing with agricultural infections and pests, such as powdery mildew or spider mites[253]. The development of UV treatment is especially important for the growing number of pathogens and pests that need chemical alternatives due to increasing fungicidal resistance. However, much more research is still needed in this field of photonics to fully comprehend all of the implications, including the most effective deployment strategies.

Despite this, the market for UV LEDs has been growing exponentially, with a five-fold expansion in the previous decade. By 2025, it is projected to surpass $1 billion[253]. To make full use of this field, UV light, when applied at the correct frequency and dosage, can enhance the amount of active compounds produced by traditional and medicinal plants while promoting a healthy growth environment. This includes boosting the antioxidant properties of various plants and increasing the levels of THC, the primary psychoactive compound in cannabis. For instance, UV-B radiation can stimulate plant responses, producing higher concentrations of flavonoids and cannabinoids[253]. Recent advancements in UV LED technology allow for the targeted application of UV light, especially UV-A and UV-B, in indoor cultivation processes. Building upon the previous point, researchers have discovered that, without UV radiation, some species of plants may develop abnormal growth patterns on leaves and shoots, resembling callus-like intumescence[253]. Standard glass, for instance, prevents more than 90% of UV-B radiation from entering, which means plants cultivated in greenhouses or similar environments without supplementary lighting may experience detrimental effects[253].

This method works because plants have inherent chemical mechanisms that allow them to recognise the various light wavelengths that trigger different specific responses. This includes the power to detect UV radiation, which can modify a plant's chemical composition and structure. Among the most significant effects UV rays have on plants is that they cause the concentration of UV-absorbing compounds, including phenolic compounds, to increase and accumulate in plants. Similar to sunscreen for humans, these UV-absorbing compounds prevent the plants from receiving damage due to overexposure to UV light. Interestingly, these phenolic compounds are not only useful to plants; studies have demonstrated them to benefit human health, including improving antioxidant properties and preventing several chronic diseases, including some types of cancers and cardiovascular diseases. For example, research has been done on the potential health benefits of resveratrol, which is present in red wine and grapes, on the heart, immune system, and even brain function[253]. The impact of UV-B radiation on the production of these compounds has been studied, with a study of rosemary demonstrating the approximate doubling of phenolics when grown under UV-B radiation. Similar conclusions were reached using Mentha spicata (spearmint), where the production of essential oils increased[253].

As mentioned, in response to the growing resistance against fungicides, UV light is needed as a possible alternative for decreasing plant pests, including but not limited to mildew and mould. This effect also seems to be linked to the UV-absorbing substances plants produce under UV radiation. These substances appear to alter the plants' "attractiveness" to these pests, which helps defend plants against infection, pests, and other harm. For indoor growers, powdery mildew poses a major threat. However, UV light has been shown to impressively reduce the growth of mildew in several plants, from cucumbers to rosemary to grapes[253]. Impressively, using appropriate doses of UV-B, researchers have managed to achieve up to 99% reduction in the severity of powdery mildew[253].

Additionally, UV-B light has been observed to aid in minimising the number of pests called survivorship mites, which can wipe out entire harvests. According to a study by Ohtsuka and Osakabe, the percentage of larvae that survived after exposure to UV-B dosages turned out to be less than 6% after two days[253]. By the third day of the experiment, all the larvae had died.

Another main type of fungus affecting plants is Botrytis cinerea, often called grey rot due to its colour, which usually affects roughly 200 distinct species of fruits or flowers[253]. Mostly, this pest enters an indoor growing area by the air or by shoes and clothing from outside. Therefore, handling this pest with full air disinfection of the room or floor disinfection systems could be required[253]. Evidence suggests that UV-C irradiation is the most effective method of treating Botrytis cinerea spores. Again, quite a high success rate was reached in 2001 when Mercier et al. used 440–2200 J/m² dosages of UV–C to obtain over 90% of disinfection rates[253].

A complete UV LED system has several other aspects to remember besides just the LEDs. These systems must also consider the necessary UV dose, the wavelength required for the specific application, and the lighting layout in relation to the plant canopy. That is not all, though: more details include adding the optical layout, the driver and the power source, thermal management, and most crucially, the lens material. With indoor gardening, it is crucial to determine the spectrum that will most effectively meet the needs of the plants. The impacts of various wavelengths vary depending on the species of plant and the plant's position in the growth cycle. For instance, a small quantity of visible green light can promote plant development. Still, studies have indicated that the benefit varies with species and that a dosage of more than 50% of green light can be harmful[253]. Thus, farmers and designers need to be clear about their goals for their plants when integrating UV light into agricultural lighting—possibly even integrating UV into the facility's main light source[253]. For example, resveratrol, the above-mentioned therapeutic molecule that plants make in response to stress, is created chemically by interacting with UV-A radiation of wavelength less than 360 nm. Similarly, to produce their desired impact, gardeners interested in raising certain flavonoid or cannabinoid levels will likely want to focus on UV-A, UV-B, or both[253].

Laser-induced breakdown spectroscopy (LIBS)

Moving on from the agricultural aspect, the food industry has many more applications for photonics throughout the whole process, from inspection to packaging to retail. Laser-induced breakdown spectroscopy (LIBS) is an atomic emission spectroscopy method that uses a laser beam to generate mycoplasma on a material's surface[254]. Explained further, the LIBS technique involves using a laser beam to interact with a material that creates plasma which emits a spectrum while it decays, reflecting the sample's elemental composition and providing spectral information about the sample's excited atoms and ions[254]. Even though plasma means the sample loses some of its molecular information, LIBS has successfully identified samples with intricate matrices. It is commonly used in the field of material sciences, but it has been proposed to be used for analysing food composition and as a part of food authentication.

Food fraud is the act of intentionally omitting or replacing an important component or ingredient of a food product. This includes economically motivated adulteration (EMA) which is the addition of substances in food to improve its appearance, flavour, or perceived worth. Dairy products are especially frequently reported to have problems, including safety issues like pathogenic microorganisms being found, incidences of fraudulent documentation, and adulteration where wood pulp is added as cellulose to prevent clumping. Therefore, high-speed food classification and authentication help ensure that fraudulent products are prevented from entering the market or removed with speed and efficiency. Other techniques for this task include vibrational spectroscopy, hyperspectral imaging, and liquid chromatography. However, these methods necessitate thorough sample preparation, expensive lab equipment, specialised technicians, and sometimes numerous chemical reagents.

Studies have investigated whether LIBS can be a more rapid alternative or supplement in food authentication. For example, verifying the authenticity of a product's protected designation of origin (PDO) is crucial in ensuring the integrity of agroalimentary products, as it guarantees their production and origin. In 2016, Moncayo et al. utilised LIBS to authenticate red wines and determine their geographic origin by analysing their unique spectral fingerprints[255]. These fingerprints are generated based on the wine's composition, which includes elements from the soil, grape variety, climate, yeasts, and winemaking processes, meaning its origins can be traced.

As mentioned, one of the main selling points of LIBS is that it does not require sample preparation. Although adding this step would increase the authentication duration, Moncayo et al. decided on a liquid-to-solid sample preparation. They would transform the liquid wine samples into gel using natural collagen and then dry them in an air-assisted oven as a novel sample preparation method[255]. Their reasoning was to enhance the analytical performance since it has been observed in other studies that lasers interacted better with solids in such scenarios with faster ablation rate, elevated plasma temperature, and increased electron density[255]. On the flip side, this method also helps avoid inherent liquid-related drawbacks like splashing and surface ripples, enabling a lower detection limit, improved repeatability, and enhanced sensitivity[255].

The combination of LIBS and machine learning has been explored and shown to be successful in many areas requiring sample classification. In this example, they tested LIBS with a neural network to accurately and efficiently discriminate between the wines[255]. The NN model underwent a three-pronged evaluation, examining its internal sensitivity, its ability to generalise, and its robustness through external testing. Breaking down each step, the network first goes through an internal validation to see if it can classify each and every wine sample used in the training. This tests its sensitivity to see if it can identify wine replicas. Then, it should demonstrate a strong capacity to generalise, meaning an ability to accurately classify wines also from one of the PDOs that weren't part of the training set[255]. This is referred to as external validation because the wine has never been seen by the model before. Finally, the robustness of the model means its ability to recognise and label wine without a designated PDO as "unassigned" rather than misclassifying it as one of the trained PDOs[255]. Ultimately, the model using LIBS as input achieved an impressive sensitivity, reaching 99.2% and remarkable generalisation ability at 98.6% while demonstrating perfect robustness, unaffected by any variations or anomalies[255].

Another example of a study would be Bilge et al., who successfully utilised LIBS to differentiate between beef, pork, and chicken meats[256]. Meat is widely consumed, but its high cost leads to limited availability and makes it an attractive target for adulteration to boost profit. One of the most common ways this adulteration is conducted is by adding less expensive meat varieties like pork or chicken to pricier ones like beef[257,258]. This type of meat adulteration has far-reaching consequences, including economic, ethical, and health issues, as well as potential violations of religious dietary laws and the risk of triggering allergic reactions that restrict the consumption of specific meat species[259].

So far, some conventional methods for detecting meat adulteration include polymerase chain reaction, gas chromatography, mass spectrometry, and high-performance liquid chromatography. The issue with these methods is that they struggle to distinguish between certain meat species due to their similar molecular compositions. Genetic methods can be used to identify meat species, and while these are dependable and highly sensitive, they have several drawbacks, which were explained earlier. Chemical methods are time-consuming as well, whereas immunological methods can be prone to inaccuracies from the cross-reactions triggered by antibody biomarkers, leading to false results[256]. Once again, Despite numerous existing methods for detecting meat adulteration, they are found to be inadequate, and the meat industry requires a more sensitive, quick, and accurate detection approach. That is why a study was conducted to test whether LIBS can better identify meat species. The distinct elemental compositions of various meat types were utilised to identify each. The LIBS spectra obtained were analysed using certain chemometric techniques. Chemometrics refers to the study of deriving useful information from chemical or biochemical systems through data-driven approaches. For instance, in this study, meat species were differentiated qualitatively utilising the Principal Component Analysis method[256]. The study yielded a determination coefficient (R^2) of 0.994 and a limit of detection (LOD) of 4.4% for beef adulterated with pork and an R^2 of 0.999 with an LOD of 2.0% for beef adulterated with chicken[256]. The R^2 value, which ranges from 0 to 1, indicates the model's predictive power, with higher values signifying better predictions. In this case, the high R^2 values suggest that the model effectively explains the variation in the outcome by using the covariates used in the model to predict the results[256].

On the other hand, the limit of detection represents the weakest signal or smallest amount of a substance that can be detected confidently with certainty, specifically when it is ensured that the signal does not originate from random fluctuations or background noise[256]. Therefore, a lower LOD is generally better as it means there can be measurements of smaller amounts of an analyte[256]. Overall, there have been promising results from studies on LIBS for food authentication, so it may become a useful tool for quality control measurements as a regular method.

Biomimetics/biomimicry

Biomimetics, also known as biomimicry, involves imitating nature's designs, processes, and structures to develop innovative solutions to complex human challenges. Specifically in photonics, there have been developments in bio-inspired photonics materials. This is slightly different from biophotonics, but it still relates to biology. Structural colour, where colour is created by the intricate nano-sized structural arrangement of materials, is a rich source of inspiration easily found in nature. Many animals, but insects in particular, have developed a wide variety of microscopic and nanoscopic structures to produce colouration, which serves purposes including but not limited to camouflage, warning off predators, marking territories, and attracting mates. Scientists have been fascinated by these structures for centuries and have intensified their research in recent decades. The study of biological photonics can be dated back to as early as ancient Greece. In the 6th century B.C.E., Anaximenes, a student of the renowned philosopher Anaximander and widely considered the first scientist, recorded the first known observation of bioluminescence. He noted a glowing effect in water when struck with an oar. Later, Aristotle also witnessed and documented this phenomenon in his works, describing encountering "things which are neither fire nor forms of fire seem to produce light by nature".

The interactions to achieve structural colour, a form of colouration, are possible because the size of these photonic structures matches the wavelength of light. These structures can manipulate light through constructive and destructive interference, enhancing certain colours while suppressing others. They are found naturally in a wide range of organisms, each with its own unique morphology tailored to achieve its desired effect. For example, a peacock's bright feathers result from its natural photonic crystals, while the vibrant blue found in certain Morpho butterflies comes from its tree-like structures[260]. These are some of the inspirations for the following innovations.

Here are some ways animals have inspired new materials. Starting off, an example of a bio-inspired photonics structure is sometimes called moth-eye anti-reflective coating[260]. Moths have unique structures within their eyes that decrease reflection rather than produce colour. This allows the moth to navigate well even in low-light conditions while preventing reflections that could reveal its location to predators. The way this structure works is that it features a honeycomb-like arrangement of tiny bumps, each standing about 200 nm tall and separated by around 300 nm[260]. This structure can serve as an anti-reflective coating because the protrusions are uniformly arranged in rows smaller than the wavelength of visible light. Light reflection occurs when the refractive indices of the air and the material the light strikes differ. However, because of these bumps' size and curved spindle shape, light can enter the eye from multiple angles while continuously changing the refractive index, enabling a smooth transition from one medium as though the light doesn't "see" the boundary[260]. With the sudden change in refractive index at the air-material interface removed, it is as though the air-lens interface is removed, resulting in more light passing through with nearly no light reflection[260].

Using this concept, we have created new kinds of practical anti-reflective coatings. For instance, Canon incorporates the moth-eye technique in their Sub-Wavelength structure Coating, resulting in a substantial decrease in lens flare[260]. The next example is something more common in daily life. There are mostly two options for improving readability under a bright light like the sun. Cell phone displays nowadays are equipped with light sensors that automatically increase screen brightness accordingly in bright environments, although this comes at the cost of battery drain. Otherwise, existing anti-reflective screens are currently available but only capture specific light wavelengths. However, according to a study published in Optica, adding an insect-inspired film to a cell phone significantly enhances screen visibility, making it four times easier to read in sunlight and an impressive ten times easier in shaded conditions[261]. Furthermore, these miniature structures are not only effective but also flexible, making them a promising candidate for coating future foldable displays. However, it may still be a while before seeing these innovative films as a common technology on mobile devices as the researchers still need to refine the flexible film to ensure it can withstand the rigours of daily use[261].

Additionally, this technology has been implemented in solar panels. One of the greatest challenges in solar power technology is the restrictions on efficiency limits. The maximum solar conversion efficiency of single junction solar cells is generally capped at around 33.7%, a limitation known as the Shockley-Queisser limit[262]. This is largely due to the inherent properties of the material used, typically silicon, as well as practical issues such as reflection. For the past few years, coating solar cells with nanomaterials has shown potential in boosting their energy absorption. Usually, these materials have taken the form of custom-designed nanowires or tiny beads. However, a team of scientists in the U.S. Department of Energy's Brooklyn National Laboratory has drawn inspiration from nature, developing a novel material modelled after the unique compound eyes of moths[262].

Their main goal was to prevent radio waves or light from bouncing at material surfaces so that more wavelengths of light could be absorbed rather than reflected. During a photovoltaic reaction, a certain amount of energy is necessary to promote an electron to transition from the valence band to the conduction band[262]. Therefore, the maximum possible efficiency can be increased if more wavelengths of the required energy can be absorbed. Although existing coatings can capture more wavelengths, they are usually restricted to one wavelength per material, with each coating optimised for light from particular directions and colours[262]. While these materials could be combined to absorb a spectrum of wavelengths, there can be possible interference. The scientists, Black and his colleagues, were searching for a better solution: a material that could absorb several wavelengths simultaneously[262].

Using the moth-eye-inspired structure, the resulting surface nanotexture created a gradual transition in refractive index on the solar panel's surface, leading to a significant reduction in the reflection of multiple wavelengths at once, improving light absorption[262]. After evaluating their novel absorbent material, Black and his team discovered that it surpassed other single-layer anti-reflective coatings by 20 percent[262]. This demonstrates that this single-material technique, which they published in the journal Nature Communications, can simplify silicon solar cells' production process, decrease overall manufacturing expenses, and accelerate the integration of solar technology[262].

Unsurprisingly, this team of scientists were not the first to be inspired by these structures for photovoltaic applications. In fact, even back in 2009, Dutch researchers had already successfully developed an anti-reflective coating inspired by the nanostructures found in a moth's eye[262]. Heat is typically lost during photovoltaic reactions, and this was the problem the group aimed to solve. Led by Jaime Gomez Rivas at the FOM Institute for Atomic and Molecular Physics in Eindhoven, they employed an environmentally friendly production method to create a coating capable of harnessing both heat and visible light to generate electricity[262].

There are numerous nanostructures found in nature to create colours and patterns. These are present in the individual strands that compose a peacock feather, known as barbules. These consist of an outer layer of keratin and an inner layer containing melanin rods linked together by keratin with gaps in between. When the melanin rods align with the keratin lattice in the outer layer, they produce a brown hue. The varying spacings of the melanin layers create the rest of the feather's colours. Humans have created structures inspired by many different kinds of natural nanostructures. For example, aperiodic photonic structures, characterised by a quasi-ordered crystal arrangement, can generate blue and green hues. There are also 3D helicoidal multilayers[260]. Imagine a stack of layers composed of fibres aligned in the same direction. In 3D helicoidal multilayers, each layer is slightly twisted or rotated relative to the one below it, causing the helical shape[260]. This unique structure enables nature to reflect polarised light, producing an intense colour effect through Bragg's reflection[260]. This phenomenon was first discovered for X-rays, but applies to any matter wave, including electron waves, as long as there is a substantial number of atoms. It occurs when light waves interact with periodic microscale structures, like the twisted layers in 3D helicoidal multilayers. When the light wave's wavelength matches the spacing between the layers, it creates a strong reflection, resulting in an intense colour effect[260]. This is similar to how a diffraction grating works but has a twisted, helical structure.

Such intricate nanostructures can be found in butterflies and beetles, which are responsible for their striking iridescence. These multilayered structures can take on various forms, not limited to only 3D but also 1D and, although more rarely, 2D.

The structures create intentional disorder and irregularity, allowing organisms to adapt to their environment. Scientists have successfully replicated these structures, which can be used as coatings to achieve stable, vibrant colours in various applications[260]. Moreover, these coatings can be designed to display specific patterns, offering a high degree of flexibility.

A specific example would be the naturally occurring 2D periodic structure in fireflies. Scientists discovered these structures when observing the light-emitting cuticles that allow fireflies to glow using scanning electron microscopy. The pattern they saw resembled a "factory roof," with scales angled diagonally and a sharp edge on the outer side, creating a sloping shape. The scientists mimicked this pattern on light-emitting diodes (LEDs) by using a special photoresist layer, leading to remarkable improvements. There was a 68% increase in power and a 55% increase in light extraction efficiency[263]. This technology means that LEDs can produce the same amount of light while using less energy, making them more energy-efficient[263].

Looking at a different approach, responsive materials can react to changes in their environment in real time. While they may require a brief moment to adapt, their behaviour mirrors the dynamic responses observed in natural systems. Examples of such mechanisms can be found in chameleons and octopuses. On the other hand, some deep-sea creatures employ a strategy called counterillumination, where they utilise a specialised luminous organ located on their underside to emit light and create the illusion of brightness from below. This organ is a laminar structure comprising phagocytes and nerve fibres, with narrow gap junctions connecting them. The interconnected network and layered architecture of neuro-phagocyte units in deep-sea fish are believed to enable their rapid, spontaneous bioluminescent responses to change situations. Since these nerve cells are directly linked to the spinal cord and brain, researchers propose that electrical signals can stimulate the phagocytes to react. Inspired by this natural phenomenon, scientists are now exploring the development of innovative technologies that incorporate these neuro-phagocyte units.

There are a plethora of possibilities for these biologically inspired materials to be applied. They can be used to create adaptive camouflage, shape-shifting devices that maintain their properties, and even biomedical innovations. For instance, a coating of this technology can facilitate the integration of foreign objects like artificial implants into living systems, such as the human body[260]. By disguising them as non-foreign entities, the antibodies don't see the object as a threat, therefore allowing a more seamless acceptance of artificial devices, like cardiac pacemakers, into the body[260].

Another interesting application is in hydrogels. Hydrogels are materials made up of hydrophilic polymers that can rapidly absorb and retain fluid, typically water. They are found in many applications, from contact lenses and cosmetics to medical applications, such as wound dressing and drug delivery. Researchers have created a hydrogel that can visually display its antibacterial and self-healing properties by leveraging biophotonics[260]. The incorporation of silver nanoparticles prevents bacteria from adhering to the surface of the hydrogel. This is important because bacterial adhesion can lead to the degradation of the hydrogel, leading to a loss of colour. Therefore, by observing the colour changes, researchers can tell if the hydrogel is damaged. If the hydrogel remains coloured or fades and then returns to its original colour, this signals successful self-healing.

Space Exploration/Development:

Expanding the scope of photonics exploration, we now venture into the realm of space exploration and development. Space environments present unique challenges and opportunities for the application of photonics technologies, particularly in terms of life support systems, health monitoring, and environmental assessment within spacecraft and extraterrestrial habitats. These aspects use very similar if not identical technology to ones mentioned already, demonstrating another significant advantage of many photonics technologies: its versatility and flexibility to be adapted for different fields. Since the technologies have already been explained, this next section will be brief simply to show how the same equipment can have different applications.

Photonics in Life Support Systems

In the constrained and isolated environments of space stations and potential extraterrestrial colonies, ensuring a stable, renewable source of life essentials such as air and water is paramount. Photonics technologies, specifically optical sensors, play a critical role in the management and recycling of these resources. Similar to the operations in agriculture, optical sensors are employed to continuously monitor the quality of air and water, detecting contaminants at molecular levels. For instance, fibre optic sensors can detect trace amounts of hazardous gases and volatile organic compounds in the air, ensuring the atmosphere within a spacecraft remains safe for astronauts. Similarly, advanced spectroscopy techniques, such as Raman spectroscopy, are utilised to assess water quality by identifying chemical pollutants and biological contaminants without the need for extensive sample preparation.

Health Monitoring in Astronauts

Maintaining astronaut health on long-duration space missions is a significant application area for biophotonics. The microgravity environment of space can lead to various health issues, including muscle atrophy, bone density loss, and alterations in vision and cardiovascular function. Biophotonic devices are integral in non-invasively monitoring the physiological changes astronauts experience. Wearable sensors embedded with photonic technologies can continuously track vital signs such as heart rate, blood oxygen levels, and arterial stiffness. Furthermore, portable diagnostic devices based on photonic biosensors enable the detection of biomarkers indicative of radiation exposure or the onset of infectious diseases, which are critical in the closed environment of a spacecraft.

Environmental Control and Monitoring:

Beyond the monitoring of life support systems, photonics plays a crucial role in the broader environmental control within spacecraft and other off-world habitats. Optical fibre sensors, capable of withstanding the harsh conditions of space, are used to monitor structural integrity, detect ionising radiation levels, and assess the spacecraft's external and internal environments. These sensors provide critical data that ensures the safety and stability of space habitats and assists in the timely maintenance and repair of spacecraft systems.

Plant Growth and Agriculture in Space:

As missions extend deeper into space and potentially to other planets, the ability to grow food locally becomes essential. Here, biophotonics intersects with space agriculture. LED-based growth systems that use specific wavelengths to optimise plant growth are vital in these settings. These systems rely on the principles of photomorphogenesis, manipulating light spectra to influence plant characteristics such as size, growth rate, and nutritional content. Moreover, hyperspectral imaging is used to monitor plant health and detect diseases early by analysing the light reflected from the plants, which varies with physiological changes.

Lunar Flashlight Satellite

The Lunar Flashlight Satellite was a small low-cost satellite whose goal was to map and identify ice in the dark regions of vast moon craters for future explorations[264]. This is part of NASA's eventual plans for "Human mission to Mars" which depends on making use of the local natural resources to make oxygen as well as propellant for launching the return ship back to Earth[264]. The flashlight emits laser light from an onboard infrared spectrometer into craters and measures the amount of reflected light to map the crater territory to identify materials like rock, dust, methane ice, and more[264]. They can also isolate specific wavelengths that reveal the amount and distribution of water discovered. Unfortunately, the Lunar Flashlight was unable to enter orbit due to a propulsion system failure and has since been abandoned[264]. Such technology has also been used on the Lunar Reconnaissance Orbiter, more specifically on the Lunar Orbiter Laser Altimeter (LOLA) which scans the lunar surface with laser beams, as well as the Mars Global Surveyor[265].

Challenges and Future Directions

While photonics offers numerous solutions for space exploration, the integration of these technologies into space missions presents unique challenges. The devices must be highly reliable, robust, and capable of operating under extreme conditions of radiation, vacuum, and temperature fluctuations. Future advancements in photonics for space exploration will likely focus on enhancing the durability and functionality of these devices, miniaturising equipment for payload efficiency, and improving the autonomy of systems to support deep-space missions where real-time communication with Earth is not feasible.

Final words

As we close this book on photonics innovations, we have journeyed through historical breakthroughs, explored the impact of biophotonics in medicine as well as other prominent fields, and glimpsed into the future of technology. From ancient marvels to modern medical advancements, the fusion of light and science has reshaped our world.

Through the lens of photonics, we have witnessed the birth of groundbreaking technologies that have not only enhanced our understanding of the universe but have also paved the way for unprecedented advancements in fields as diverse as healthcare, telecommunications, and environmental sustainability. The chapter on biophotonics in medicine has shown how light is transforming healthcare, offering non-invasive diagnostics and targeted treatments. The following chapter covers some other fields of interest for photonics developments but it simply touches upon the surface; it is nowhere near an accurate representation of the true range of possibilities offered by photonics. Even the boundaries of medicine and especially space exploration continue to expand. As you delve further into photonics, its versatility becomes apparent as the same innovations can be applied to several areas of interest such as LiDAR being applicable to autonomous driving, environmental monitoring, infrastructure applications, etc. The global impact is further emphasized by the potential in applications like renewable energy. Looking ahead, emerging technologies like quantum photonics, artificial intelligence, and photonics integrated circuits promise further advancements in computing and energy. Even within the short span of the book, it is already shown that photonics has had a dominating impact in a multitude of fields.

In wrapping up, let us carry forward the lessons learned and the potential of photonics to drive innovation. Let's continue to harness the power of light to shape a future where technology improves lives and pushes the boundaries of what's possible.

References

1 *Photoelectric effect* (2024) *Encyclopædia Britannica*. Available at: https://www.britannica.com/science/photoelectric-effect (Accessed: 21 October 2024).

2 *History of Quantum Mechanics* (2024) *Wikipedia*. Available at: https://en.wikipedia.org/wiki/History_of_quantum_mechanics (Accessed: 21 October 2024).

3 Dobrijevic, D. (2022) *The double-slit experiment: Is light a wave or a particle?*, *Space.com*. Available at: https://www.space.com/double-slit-experiment-light-wave-or-particle (Accessed: 21 October 2024).

4 Marianne (1970) *Physics in a minute: The double slit experiment*, *Plus Maths*. Available at: https://plus.maths.org/content/physics-minute-double-slit-experiment-0 (Accessed: 21 October 2024).

5 *Photon* (2024) *Wikipedia*. Available at: https://en.wikipedia.org/wiki/Photon (Accessed: 21 October 2024).

6 Jones, A.Z. (2018) *What exactly is a photon?*, *ThoughtCo*. Available at: https://www.thoughtco.com/what-is-a-photon-definition-and-properties-2699039 (Accessed: 21 October 2024).

7 Townes, C. and Schawlow, A. (2003) *This Month in physics history | american physical society*, *APS125*. Available at: https://www.aps.org/apsnews/2003/12/this-month-in-physics-history (Accessed: 21 October 2024).

8 *Historic trends in light intensity* (2020) *AI Impacts*. Available at: https://aiimpacts.org/historic-trends-in-light-intensity/ (Accessed: 21 October 2024).

9 *The first laser* (no date) *The University of Chicago Press*. Available at: https://press.uchicago.edu/Misc/Chicago/284158_townes.html (Accessed: 21 October 2024).

10 Thompson, J. (2024, March 30). How do lasers work? *Livescience*. Available at: https://www.livescience.com/physics-mathematics/how-do-lasers-work (Accessed: 21 October 2024).

11 Weschler, M. (2023) *How lasers work*, HowStuffWorks Science. Available at: https://science.howstuffworks.com/laser.htm (Accessed: 21 October 2024).

12 *NIF's guide to how lasers work* (no date) *National Ignition Facility & Photon Science*. Available at: https://lasers.llnl.gov/education/how-lasers-work (Accessed: 21 October 2024).

13 Gallegos, J. (2023) *Refractive index, StatPearls [Internet]*. Available at: https://www.ncbi.nlm.nih.gov/books/NBK592413/ (Accessed: 21 October 2024).

14 Lalande, A. (2020) *How does light carry data across optical fiber?*, *Medium*. Available at: https://medium.com/@BillyBBone/how-does-light-carry-data-across-optical- fiber-783740c384d8 (Accessed: 21 October 2024).

15 *Photonics* (2024) *Wikipedia*. Available at: https://en.wikipedia.org/wiki/Photonics (Accessed: 21 October 2024).

16 *What is Integrated Photonics* (no date) *AIM Photonics*. Available at: https://www.aimphotonics.com/what-is-integrated-photonics (Accessed: 21 October 2024).

17 Moore's law (2024) Wikipedia. Available at: https://en.wikipedia.org/wiki/Moore's_law (Accessed: 21 October 2024).

18 Hajj, A.E. (2022) *What reaching the size limit of the transistor means for the future*, *Inside Telecom*. Available at: https://insidetelecom.com/what-reaching-the-size-limit-of-the-transistor-means-for-the-future/ (Accessed: 21 October 2024).

19 Laskey, R. (2022) *What is a photonic integrated circuit? - explained by Photondelta*, *PhotonDelta*. Available at: https://www.photondelta.com/news/what-is-a-photonic-integrated-circuit/ (Accessed: 21 October 2024).

20 Yamashita, P. (2023) *Rockley photonics advances non-invasive blood glucose monitoring*, *Rockley Photonics*. Available at: https://rockleyphotonics.com/2023/09/26/rockley-photonics-advances-non-invasive-blood-glucose-monitoring/ (Accessed: 21 October 2024).

21 Shahbaz, M., Butt, M.A. and Piramidowicz, R. (2023) Breakthrough in Silicon Photonics Technology in telecommunications, biosensing, and Gas Sensing, MDPI. Available at: https://www.mdpi.com/2072-666X/14/8/1637 (Accessed: 21 October

2024).

22 *Integrated Quantum Photonics* (2024) *Wikipedia*. Available at: https://en.wikipedia.org/wiki/Integrated_quantum_photonics (Accessed: 21 October 2024).

23 Giles, M. (2024) *Explainer: What is a quantum computer?*, *MIT Technology Review*. Available at: https://www.technologyreview.com/2019/01/29/66141/what-is-quantum-computing/ (Accessed: 21 October 2024).

24 *Nanophotonics* (2024) *Wikipedia*. Available at: https://en.wikipedia.org/wiki/Nanophotonics (Accessed: 21 October 2024).

25 *Dennis Gabor* (2004) *Molecular expressions*. Available at: https://micro.magnet.fsu.edu/optics/timeline/people/gabor.html (Accessed: 21 October 2024).

26 Beléndez, A. (2004) *Dennis Gabor, "father of holography" - openmind copia - rua*, *Open mind*. Available at: https://rua.ua.es/dspace/bitstream/10045/48305/1/Dennis-Gabor-OpenMind-05-06-2015-ENG.pdf (Accessed: 21 October 2024).

27 *Emmett Leith* (2016) *MICL*. Available at: https://micl.engin.umich.edu/stories/emmett-leith (Accessed: 21 October 2024).

28 Leith, E.N. (2007) *A short history of the Optics Group of the willow run laboratories*, *Science Direct*. Available at: https://www.sciencedirect.com/science/article/pii/B9780121860301500035 (Accessed: 21 October 2024).

29 (No date) *Chapter 10: Off-Axis "Leith & Upatnieks" Holograms* . Available at: https://ocw.mit.edu/courses/mas-450-holographic-imaging-spring-2003/5e7cf4fa600d2c2e1cf06a6e735791e9_ch10offaxisluholos.pdf (Accessed: 21 October 2024).

30 *History of Fiber Optics* (2020) *Timbercon*. Available at: https://www.timbercon.com/resources/blog/history-of-fiber-optics/ (Accessed: 21 October 2024).

31 The French solar power pioneer who was Light Years Ahead (2023) The

Guardian. Available at: https://www.theguardian.com/environment/2023/jul/27/augustin-mouchot-french-pioneer-solar-power-engine (Accessed: 21 October 2024).

32 The history of solar - IIS windows server. Available at: https://www1.eere.energy.gov/solar//pdfs/solar_timeline.pdf (Accessed: 21 October 2024).

33 Richardson, L. (2023) Solar history: Timeline & invention of solar panels, EnergySage. Available at: https://www.energysage.com/about-clean-energy/solar/the-history-and-invention-of-solar-panel-technology/ (Accessed: 21 October 2024).

34 History (2020) National Candle Association. Available at: https://candles.org/history/ (Accessed: 21 October 2024).

35 Staff, S.L. (2016) The historical evolution of lighting, LED Lighting Distributor and Implementation Company. Available at: https://www.stouchlighting.com/blog/the-historical-evolution-of-lighting (Accessed: 21 October 2024).

36 Solid-state lighting (2024) Wikipedia. Available at: https://en.wikipedia.org/wiki/Solid-state_lighting (Accessed: 21 October 2024).

37 History of LED lighting (no date) Shine Retrofits Lighting Blog. Available at: https://www.shineretrofits.com/lighting-center/lighting-resources/history-of-led-lighting/ (Accessed: 21 October 2024).

38 Blue Leds – filling the world with new light. Available at: https://www.nobelprize.org/uploads/2018/06/popular-physicsprize2014-1.pdf (Accessed: 21 October 2024).

39 Courtland, R. (2021) Inventors of Blue led win Nobel prize in physics, IEEE Spectrum. Available at: https://spectrum.ieee.org/inventors-of-blue-led-win-nobel-prize-in-physics (Accessed: 21 October 2024).

40 Diep, F. (2014) Why a blue led is worth A nobel prize, Popular Science. Available at: https://www.popsci.com/article/technology/why-blue-led-worth-nobel-prize/ (Accessed: 21 October 2024).

41 Eugen (2023) When LED lights were invented? A brief history of LED lighting - led & lighting info, LED & Lighting Info - Useful Tips To Improve Design. Available at: https://ledlightinginfo.com/led-lights-history (Accessed: 21 October 2024).

42 Hilton, S. (2019). Silver Nanowires make Transparent and Flexible LED Screens. [online] TEAM-TRADE Blog. Available at: https://blog.teamtrade.cz/silver-nanowires-make-transparent-and-flexible-led-screens/ [Accessed 26 Oct. 2024].

43 Negative-index metamaterial (2024) Wikipedia. Available at: https://en.wikipedia.org/wiki/Negative-index_metamaterial (Accessed: 21 October 2024).

44 History of metamaterials (no date) Wikiwand. Available at: https://www.wikiwand.com/en/articles/History_of_metamaterials (Accessed: 21 October 2024).

45 Metamaterials: History (no date) David R. Smith Group. Available at: https://people.ee.duke.edu/~drsmith/metamaterials/metamaterials_history.htm (Accessed: 21 October 2024).

46 Superlens (2024) Wikipedia. Available at: https://en.wikipedia.org/wiki/Superlens (Accessed: 21 October 2024).

47 Avril, T. (2018) Eli Yablonovitch, Eli Yablonovitch | Research UC Berkeley. Available at: https://vcresearch.berkeley.edu/faculty/eli-yablonovitch (Accessed: 21 October 2024).

48 Photonic Crystal (2024) Wikipedia. Available at: https://en.wikipedia.org/wiki/Photonic_crystal (Accessed: 21 October 2024).

49 Shanker, R. (2015) Self-assembled photonic crystals infiltrated with nanoplatelets and nanotubes, Research portal. Available at: http://epubs.surrey.ac.uk/808023/ (Accessed: 21 October 2024).

50 Admin, C.W. (2023) Eli Yablonovitch: Father of Photonic Band Engineering, Berkeley Engineering. Available at: https://engineering.berkeley.edu/eli-yablonovitch-father-of-photonic-band-engineering/ (Accessed: 21 October 2024).

51 Eli Yablonovitch (no date) Eli Yablonovitch | EECS at UC Berkeley. Available

at: https://www2.eecs.berkeley.edu/Faculty/Homepages/yablonovitch.html (Accessed: 21 October 2024).

52 Huang, L.-D. (2022) Brighten the future: Photobiomodulation and Optogenetics, Focus (American Psychiatric Publishing). Available at: https://www.ncbi.nlm.nih.gov/pmc/articles/PMC9063588/ (Accessed: 21 October 2024).

53 Optogenetics (2024) Wikipedia. Available at: https://en.wikipedia.org/wiki/Optogenetics (Accessed: 21 October 2024).

54 Joshi, J., Rubart, M. and Zhu, W. (2020) *Optogenetics: Background, methodological advances and potential applications for cardiovascular research and medicine, Frontiers in bioengineering and biotechnology.* Available at: https://www.ncbi.nlm.nih.gov/pmc/articles/PMC7000355/ (Accessed: 21 October 2024).

55 (No date) *Optogenetics: Shedding light on the brain's secrets.* Available at: https://www.scientifica.uk.com/learning-zone/optogenetics-shedding-light-on-the-brains-secrets (Accessed: 21 October 2024).

56 Joshi, J., Rubart, M. and Zhu, W. (2019) *Optogenetics: Background, methodological advances and potential applications for cardiovascular research and medicine, Frontiers.* Available at: https://www.frontiersin.org/journals/bioengineering-and-biotechnology/articles/10.3389/fbioe.2019.00466/full (Accessed: 21 October 2024).

57 Boyden, E.S. et al. (2005) *Millisecond-timescale, genetically targeted optical control of neural activity, Nature neuroscience.* Available at: https://pubmed.ncbi.nlm.nih.gov/16116447/ (Accessed: 21 October 2024).

58 Dobrijevic, D. (2022) *The double-slit experiment: Is light a wave or a particle?, Space.com.* Available at: https://www.space.com/double-slit-experiment-light-wave-or-particle (Accessed: 21 October 2024).

59 Vlasits, A. (2023) *He may have invented one of Neuroscience's biggest advances. but you've never heard of him, STAT.* Available at: https://www.statnews.com/2016/09/01/optogenetics/ (Accessed: 21 October 2024).

60 Vlasits, A. (2024) *He may have invented one of Neuroscience's biggest*

advances--but you've never heard of him, Scientific American. Available at: https://www.scientificamerican.com/article/he-may-have-invented-one-of-neuroscience-s-biggest-advances-but-you-ve-never-heard-of-him/ (Accessed: 21 October 2024).

61 Boyden, E.S. (2015) *Optogenetics and the future of Neuroscience, Nature News.* Available at: https://www.nature.com/articles/nn.4094 (Accessed: 21 October 2024).

62 *History of Quantum Mechanics* (2024) *Wikipedia.* Available at: https://en.wikipedia.org/wiki/History_of_quantum_mechanics (Accessed: 21 October 2024).

63 Carson, C. (no date) *The origins of the quantum theory.* Available at: https://www.slac.stanford.edu/pubs/beamline/30/2/30-2-carson.pdf (Accessed: 21 October 2024).

64 Resonance (2024) *The history of quantum computing you need to know [2024], The Quantum Insider.* Available at: https://thequantuminsider.com/2020/05/26/history-of-quantum-computing/ (Accessed: 21 October 2024).

65 Chandrakant, K. (2024) *Introduction to quantum computing, Baeldung on Computer Science.* Available at: https://www.baeldung.com/cs/quantum-computing (Accessed: 21 October 2024).

66 Garisto, D. (2024b) *Light-based quantum computer exceeds fastest classical supercomputers*, Scientific American. Available at: https://www.scientificamerican.com/article/light-based-quantum-computer-exceeds-fastest-classical-supercomputers/ (Accessed: 21 October 2024).

67 Pednault, E. *et al.* (no date) *On 'Quantum supremacy'*, IBM Quantum Computing Blog. Available at: https://www.ibm.com/quantum/blog/on-quantum-supremacy (Accessed: 21 October 2024).

68 Castelluccio, M. (2020) *Quantum Computing in 2021*, IMA. Available at: https://www.sfmagazine.com/technotes/2020/december/quantum-computing-in-2021/ (Accessed: 21 October 2024).

69 Roger (2020) *Quantum supremacy claimed again, Quantum Supremacy claimed again.* Available at: http://blog.darkbuzz.com/2020/12/quantum-supremacy-

claimed-again.html (Accessed: 21 October 2024).

70 User, S. (no date) *A fascinating historical time line on lenses*, *Historical Timeline on Lenses*. Available at: https://www.windhoek-optics.com/index.php/2015-09-03-10-59-08/historical-timeline-on-lenses (Accessed: 21 October 2024).

71 Kellermann, K.I. and Klock, B.K. (2024) *Telescope*, *Encyclopædia Britannica*. Available at: https://www.britannica.com/science/optical-telescope (Accessed: 21 October 2024).

72 *Prism* (no date) *Encyclopædia Britannica*. Available at: https://www.britannica.com/technology/prism-optics (Accessed: 21 October 2024).

73 *Introduction to Optical Prisms* (no date) *GoPhotonics*. Available at: https://www.gophotonics.com/whitepapers/details/137-introduction-to-optical-prisms (Accessed: 21 October 2024).

74 *Introduction to optical prisms | edmund optics*. Available at: https://www.edmundoptics.com/knowledge-center/application-notes/optics/introduction-to-optical-prisms/ (Accessed: 21 October 2024).

75 Zee, E. van and Gire, E. (2020) *Ix. historical and current perspectives on the nature of light*, *Exploring Physical Phenomena*. Available at: https://open.oregonstate.education/physicsforteachers/chapter/historical-and-current-perspectives-on-the-nature-of-light/ (Accessed: 21 October 2024).

76 Lazarus, R. (2021) *What are prism lenses?*, *Optometrists Network*. Available at: https://www.optometrists.org/vision-therapy/neuro-optometry/what-are-prism-lenses/ (Accessed: 21 October 2024).

77 Porter, D. (2023) *What is Prism Correction in eyeglasses?*, *American Academy of Ophthalmology*. Available at: https://www.aao.org/eye-health/glasses-contacts/what-is-prism-correction-in-eyeglasses (Accessed: 21 October 2024).

78 *Phase* (no date) *Encyclopædia Britannica*. Available at: https://www.britannica.com/science/phase-mechanics (Accessed: 21 October 2024).

79 *Polarization of light - definition, types, methods, & applications* (2023) *BYJUS*. Available at: https://byjus.com/physics/polarization-of-light/ (Accessed: 21 October 2024).

80 *Introduction to polarization | edmund optics* (no date) *EO Edmund optics|worldwide*. Available at: https://www.edmundoptics.com/knowledge-center/application-notes/optics/introduction-to-polarization/ (Accessed: 21 October 2024).

81 Packard, K.S. (1984) *The origin of waveguides: A case of multiple rediscovery | IEEE Journals & Magazine | IEEE Xplore*, IEEE Xplore. Available at: https://ieeexplore.ieee.org/document/1132809 (Accessed: 21 October 2024).

82 Paschotta, R. (2024) Channel waveguides, – strip waveguides, fabrication, semiconductor, dielectric, waveguide properties, propagation loss, bend loss, applications. Available at: https://www.rp-photonics.com/channel_waveguides.html (Accessed: 21 October 2024).

83 Surmenev, R.A. and Surmeneva, M.A. (2023) Photodetector, Photodetector - an overview | ScienceDirect Topics. Available at: https://www.sciencedirect.com/topics/engineering/photodetector (Accessed: 21 October 2024).

84 Diode (2024) Wikipedia. Available at: https://en.wikipedia.org/wiki/Diode (Accessed: 21 October 2024).

85 What is a photonic integrated circuit (PIC) and how does it work? (no date) Synopsys. Available at: https://www.synopsys.com/glossary/what-is-a-photonic-integrated-circuit.html (Accessed: 21 October 2024).

86 Paschotta, R. (n.d.). Photonic Integrated Circuits. [online] RP Photonics Encyclopedia. Available at: https://www.rp-photonics.com/photonic_integrated_circuits.html [Accessed 26 Oct. 2024].

87 Photonic integrated circuits (PICS) for Next Generation Space Applications NASA electronics parts and packaging (NEPP) (no date) Docslib. Available at: https://docslib.org/doc/6196424/photonic-integrated-circuits-pics-for-next-generation-space- applications-nasa-electronics-parts-and-packaging-nepp (Accessed: 21 October 2024).

88 Glytsis, E.N. (no date) Introduction & evolution of integrated optics & applications. Available at: http://users.ntua.gr/eglytsis/IO/General_Information_p.pdf (Accessed: 21 October 2024).

89 van Gerven, P. (2023). Integrated photonics platforms compared: SiN, InP and SiPh. [online] PhotonDelta. Available at: https://www.photondelta.com/news/integrated-photonics-platforms-compared-silicon-nitride-indium-phosphide-silicon-photonics/ [Accessed 26 Oct. 2024].

90 Physics LibreTexts. (n.d.). 3.5: Interference in Thin Films. [online] Available at: https://phys.libretexts.org/Bookshelves/University_Physics/University_Physics_(OpenStax)/University_Physics_III_-_Optics_and_Modern_Physics_(OpenStax)/03:_Interference/3.05:_Interference_in_Thin_Films [Accessed 26 Oct. 2024].

91 Garcia, D. (2018) How mechanical switches work, Dygma. Available at: https://dygma.com/blogs/stories/how-mechanical-switches-work (Accessed: 21 October 2024).

92 PageFly (2024) What is Optical Switch and why choose it?, Keychron. Available at: https://www.keychron.com/blogs/news/what-is-optical-switch-and-why-choose-it-1 (Accessed: 21 October 2024).

93 Admin (2023) Optical switches overview, Fiber Optic Components. Available at: https://www.fiber-optic-components.com/optical-switches-overview.html (Accessed: 21 October 2024).

94 Coupling (electronics) (2024) Wikipedia. Available at: https://en.wikipedia.org/wiki/Coupling_(electronics) (Accessed: 21 October 2024).

95 Optical Coupler (no date) Optical Coupler - an overview | ScienceDirect Topics. Available at: https://www.sciencedirect.com/topics/engineering/optical-coupler (Accessed: 21 October 2024).

96 What is electrical isolation? (2022) Atlas Scientific. Available at: https://atlas-scientific.com/blog/what-is-electrical-isolation/ (Accessed: 21 October 2024).

97 Optical fiber (2024) Wikipedia. Available at: https://en.wikipedia.org/wiki/Optical_fiber (Accessed: 25 October 2024).

98 John (2024) What is an optical splitter?: FS Community, Knowledge. Available at: https://community.fs.com/article/what-is-a-fiber-optic-splitter-2.html (Accessed: 25 October 2024).

99 What is a Passive Optical Network (PON)?: Juniper Networks Us (no date) Juniper Networks. Available at: https://www.juniper.net/us/en/research-topics/what-is-pon.html (Accessed: 25 October 2024).

100 (No date) Various optical amplifiers (EDFA, FRA, and soa) | anritsu asia pacific. Available at: https://www.anritsu.com/en-au/sensing-devices/guide/optical-amplifier (Accessed: 25 October 2024).

101 Biophotonics articles from across Nature Portfolio (no date) Nature news. Available at: https://www.nature.com/subjects/biophotonics (Accessed: 25 October 2024).

102 Janith, G.I. et al. (2023) 'Advances in surface plasmon resonance biosensors for Medical Diagnostics: An overview of recent developments and techniques', Journal of Pharmaceutical and Biomedical Analysis Open, 2, p. 100019. doi:10.1016/j.jpbao.2023.100019.

103 Graybeal, J.D. and Stoner, J.O. (2024) Spectroscopy, Encyclopædia Britannica. Available at: https://www.britannica.com/science/spectroscopy (Accessed: 25 October 2024).

104 Atascientific (2020) Understanding spectrometry and Spectroscopy, ATA Scientific. Available at: https://www.atascientific.com.au/spectrometry/ (Accessed: 25 October 2024).

105 Spectroscopy: Reading the rainbow | hubblesite (2022) NASA HUBBLESITE. Available at: https://hubblesite.org/contents/articles/spectroscopy-reading-the-rainbow (Accessed: 25 October 2024).

106 MRI (Magnetic Resonance Imaging) (2018) Magnetic Resonance Spectroscopy, MR spectroscopy. Available at: https://mayfieldclinic.com/pe-mrspectroscopy.htm (Accessed: 25 October 2024).

107 About MRI " the merritt laboratory " College of Medicine " University of Florida (no date) UF monogram. Available at: https://merritt.biochem.med.ufl.edu/about-mri/ (Accessed: 25 October 2024).

108 Health topics - medical tests: Mayfield Clinic (no date) Health topics - Medical tests | Mayfield Brain & Spine. Available at: https://mayfieldclinic.com/ht_medtests.htm (Accessed: 25 October 2024).

109 MRI (Magnetic Resonance Imaging) (2018) Magnetic Resonance Spectroscopy, MR spectroscopy. Available at: https://mayfieldclinic.com/pe-mrspectroscopy.htm (Accessed: 25 October 2024).

110 (2006) MRI uses fundamental physics for clinical diagnosis | american physical society. Available at: https://www.aps.org/apsnews/2006/07/mri-fundamental-clinical-diagnosis (Accessed: 25 October 2024).

111 Garet, L. (2023) History of mris and the evolution of this life-saving technology, Ezra. Available at: https://ezra.com/blog/history-of-mri-scans (Accessed: 25 October 2024).

112 Rigden, J. (2019) Isidor Isaac Rabi: Walking the path of god, Physics World. Available at: https://physicsworld.com/a/isidor-isaac-rabi-walking-the-path-of-god/ (Accessed: 25 October 2024).

113 Stanley, Jeffrey A and Raz, N. (2018) Functional magnetic resonance spectroscopy: The 'new' Mrs for cognitive neuroscience and psychiatry research, Frontiers in psychiatry. Available at: https://www.ncbi.nlm.nih.gov/pmc/articles/PMC5857528/ (Accessed: 25 October 2024).

114 Kariuki, C. (2022) An overview of modern brain-imaging techniques - icjs - international collegiate journal of science, ICJS. Available at: https://icjs.us/an-overview-of-modern-brain-imaging-techniques/ (Accessed: 25 October 2024).

115 Théau, J. (1970) Temporal resolution, SpringerLink. Available at: https://link.springer.com/referenceworkentry/10.1007/978-0-387-35973-1_1376 (Accessed: 25 October 2024).

116 Gevins, A. et al. (1995) 'Mapping cognitive brain function with modern high-resolution electroencephalography', Trends in Neurosciences, 18(10), pp. 429–436. doi:10.1016/0166-2236(95)94489-r.

117 Stanley, Jeffrey A. and Raz, N. (2018) Functional magnetic resonance spectroscopy: The 'new' Mrs for cognitive neuroscience and psychiatry research, Frontiers. Available at: https://www.frontiersin.org/journals/psychiatry/articles/10.3389/fpsyt.2018.00076/full (Accessed: 25 October 2024).

118 Yong, E. (2009) Photographing the glow of the human body, Science. Available at: https://www.nationalgeographic.com/science/article/photographing-the-glow-of-the-human-body (Accessed: 25 October 2024).

119 yong, ed (2009) Edyong, ScienceBlogs. Available at: https://scienceblogs.com/notrocketscience/2009/07/20/photographing-the-glow-of-the-human-body (Accessed: 25 October 2024).

120 Barth, H. and Glasser, O. (1939) 'Further studies on the problem of Mitogenetic radiation', Radiology, 33(1), pp. 25–33. doi:10.1148/33.1.25.

121 Volodyaev, I. and Naumova, E.V. (2023) 'Mitogenetic Rays', Ultra-Weak Photon Emission from Biological Systems, pp. 7–26. doi:10.1007/978-3-031-39078-4_2.

122 SF, A. (no date) Dr. Fritz Albert Popp: Biophotons, Infopathy. Available at: https://www.infopathy.com/en/posts/dr-fritz-albert-popp-biophotons (Accessed: 25 October 2024).

123 Dr. Voll: Electro acupuncture voll: Energetic remedy testing: Biophoton Services California (2015) Biophoton Services California -. Available at: https://biophotonservices.com/electroacupuncture-according-to-voll/ (Accessed: 25 October 2024).

124 Chiren® 3.0 (no date) Health Angel Foundation. Available at: http://www.biontology.com/biophoton-instruments/chiren-2/ (Accessed: 25 October 2024).

125 Yong, E. (2009) Photographing the glow of the human body, Science. Available at: https://www.nationalgeographic.com/science/article/photographing-the-glow-of-the-human-body (Accessed: 25 October 2024).

126 Smotrys, M.A. et al. (2023) 'Energetic homeostasis achieved through biophoton energy and accompanying medication treatment resulted in sustained levels of thyroiditis-Hashimoto's, Iron, vitamin D & Vitamin B12', Metabolism Open, 18, p. 100248. doi:10.1016/j.metop.2023.100248.

127 Fluorescence (2024) Wikipedia. Available at: https://en.wikipedia.org/wiki/Fluorescence (Accessed: 25 October 2024).

128 Jensen, E.C. (2012) 'Use of fluorescent probes: Their effect on cell biology and limitations', The Anatomical Record, 295(12), pp. 2031–2036. doi:10.1002/ar.22602.

129 Ehrenberg, M., 2008. The green fluorescent protein: discovery, expression and development. Information Department, The Royal Swedish Academy of Sciences.

130 Green fluorescent protein (2024) Wikipedia. Available at: https://en.wikipedia.org/wiki/Green_fluorescent_protein (Accessed: 25 October 2024).

131 MCherry (2024) Wikipedia. Available at: https://en.wikipedia.org/wiki/MCherry (Accessed: 25 October 2024).

132 Bassiri, E. (no date) Enumeration of microorganisms I. objectives. Available at: https://www.sas.upenn.edu/LabManuals/biol275/Table_of_Contents_files/14-Enumeration.pdf (Accessed: 25 October 2024).

133 Plate count (no date) Plate Count - an overview | ScienceDirect Topics. Available at: https://www.sciencedirect.com/topics/immunology-and-microbiology/plate-count (Accessed: 25 October 2024).

134 MCherry fluorescent protein (no date) Takara. Available at: https://www.takarabio.com/products/gene-function/fluorescent-proteins/fluorescent-protein-plasmids/red-fluorescent-proteins/mcherry-fluorescent-protein (Accessed: 25 October 2024).

135 Refaat, A. et al. (2022) In vivo fluorescence imaging: Success in preclinical imaging paves the way for clinical applications, Journal of nanobiotechnology. Available at: https://www.ncbi.nlm.nih.gov/pmc/articles/PMC9571426/ (Accessed: 25 October 2024).

136 How to make cells glow (2022) Institute of Science and Technology Austria (ISTA). Available at: https://ist.ac.at/en/news/how-to-make-cells-glow/ (Accessed: 25 October 2024).

137 Refaat, A. et al. (2022) 'In vivo fluorescence imaging: Success in preclinical imaging paves the way for clinical applications', Journal of Nanobiotechnology, 20(1). doi:10.1186/s12951-022-01648-7.

138 Fluorescent transgenic zebrafish: How are they used in drug screening? (no date) Blog. Available at: https://blog.biobide.com/fluorescent-transgenic-zebrafish-how-are-they-used-in-drug-screening (Accessed: 26 October 2024).

139 Biobide (n.d.). About Us. [online] Biobide. Available at: https://biobide.com/about-us [Accessed 26 Oct. 2024].

140 Biobide (2024) Decoding mcherry: Advancements in fluorescent proteins, Biobide. Available at: https://biobide.com/mcherry-properties-and-applications (Accessed: 26 October 2024).

141 Ecotoxicity testing methods and models (no date) Blog. Available at: https://blog.biobide.com/ecotoxicity-testing-methods-and-models (Accessed: 26 October 2024).

142 Cleveland clinic (n.d.). Flow Cytometry: Test, Use, Analysis & Results Interpretation. [online] Cleveland Clinic. Available at: https://my.clevelandclinic.org/health/diagnostics/22086-flow-cytometry [Accessed 26 Oct. 2024].

143 Thermo Fisher Scientific. (2020). How a Flow Cytometer Works | Thermo Fisher Scientific - UK. [online] Available at: https://www.thermofisher.com/us/en/home/life-science/cell-analysis/cell-analysis-learning-center/molecular-probes-school-of-fluorescence/flow-cytometry-basics/flow-cytometry-fundamentals/how-flow-cytometer-works.html [Accessed 26 Oct. 2024].

144 McKinnon, K.M. (2018). Flow Cytometry: an Overview. Current Protocols in Immunology, [online] 120(1), pp.5.1.1–5.1.11. doi:https://doi.org/10.1002/cpim.40.

145 Barteneva, N.S., Fasler-Kan, E. and Vorobjev, I.A. (2012). Imaging Flow Cytometry. Journal of Histochemistry & Cytochemistry, [online] 60(10), pp.723–733. doi:https://doi.org/10.1369/0022155412453052.

146 Yoon, S.A. et al. (2021b) Strategies of detecting bacteria using fluorescence-based dyes, Frontiers in chemistry. Available at: https://www.ncbi.nlm.nih.gov/pmc/articles/PMC8397417/ (Accessed: 26 October 2024).

147 Silhavy, T.J., Kahne, D. and Walker, S. (2010b). The bacterial cell envelope.

Cold Spring Harbor Perspectives in Biology, [online] 2(5). doi:https://doi.org/10.1101/cshperspect.a000414.

148 Yoon, S.A. et al. (2021a) Strategies of detecting bacteria using fluorescence-based dyes, Frontiers. Available at: https://www.frontiersin.org/journals/chemistry/articles/10.3389/fchem.2021.743923/full (Accessed: 26 October 2024).

149 Martín-Yerga, D., González-García, M.B. and Costa-García, A. (2013). Electrochemical determination of mercury: A review. Talanta, [online] 116, pp.1091–1104. doi:https://doi.org/10.1016/j.talanta.2013.07.056.

150 Parks, J.M., Johs, A., Podar, M., Bridou, R., Hurt, R.A., Smith, S.D., Tomanicek, S.J., Qian, Y., Brown, S.D., Brandt, C.C., Palumbo, A.V., Smith, J.C., Wall, J.D., Elias, D.A. and Liang, L. (2013). The Genetic Basis for Bacterial Mercury Methylation. Science, [online] 339(6125), pp.1332–1335. doi:https://doi.org/10.1126/science.1230667.

151 Roeßler, M., Sewald, X. and Müller, V. (2003). Chloride dependence of growth in bacteria. FEMS Microbiology Letters, [online] 225(1), pp.161–165. doi:https://doi.org/10.1016/s0378-1097(03)00509-3.

152 The Microscope (2019) Science Museum. Available at: https://www.sciencemuseum.org.uk/objects-and-stories/medicine/microscope (Accessed: 26 October 2024).

153 Rice, G. (2024) Fluorescent microscopy, Microscopy. Available at: https://serc.carleton.edu/microbelife/research_methods/microscopy/fluromic.html (Accessed: 26 October 2024).

154 The fluorescence microscope - preparation of specimen (no date) educationalgames.nobelprize.org. Available at: https://educationalgames.nobelprize.org/educational/physics/microscopes/fluorescence/preparation.html (Accessed: 26 October 2024).

155 Desk, E. (2023) Light microscopy vs electron microscopy – understanding the differences and applications, BiotechReality. Available at: https://www.biotechreality.com/2023/04/light-microscopy-vs-electron-microscopy-understanding-the-differences-and-applications.html (Accessed: 26 October 2024).

156 Koenig, F. (2020a) Fluorescence microscopy - explanation and labelled images, New York Microscope Company. Available at: https://microscopeinternational.com/fluorescence-microscopy/ (Accessed: 26 October 2024).

157 Mondal, D.D. et al. (2023) 'An overview of nutritional profiling in foods: Bioanalytical techniques and useful protocols', Frontiers in Nutrition, 10. doi:10.3389/fnut.2023.1124409.

158 Introduction to fluorescence microscopy (2024) JoVE. Available at: https://www.jove.com/v/5040/fluorescence-microscopy-concept-instrumentation-and- applications (Accessed: 26 October 2024).

159 Koenig, F. (2020b) Fluorescence microscopy vs. light microscopy, New York Microscope Company. Available at: https://microscopeinternational.com/fluorescence-vs-light-microscopy/ (Accessed: 26 October 2024).

160 Alves, F. and Brahme, A. (2014) Comprehensive biomedical physics. 4, Optical Molecular Imaging. Amsterdam: Elsevier.

161 Worsfold, P.J., Townshend, A. and Poole, C.F. (1995) Encyclopedia of analytical science. London: Academic Press.

162 Park, S. et al. (2018) 'Superresolution fluorescence microscopy for 3D reconstruction of thick samples', Molecular Brain, 11(1). doi:10.1186/s13041-018-0361-z.

163 Lee, J. et al. (2017) 'Accelerated super-resolution imaging with fret-paint', Molecular Brain, 10(1). doi:10.1186/s13041-017-0344-5.

164 Grienberger, C. et al. (2022) 'Two-photon calcium imaging of neuronal activity', Nature Reviews Methods Primers, 2(1). doi:10.1038/s43586-022-00147-1.

165 Boston University Neurophotonics Center. (n.d.). Laser Speckle Contrast Imaging | Neurophotonics Center. [online] Available at: https://www.bu.edu/neurophotonics/research/lsci/ [Accessed 27 Oct. 2024].

166 Yuan, S., Devor, A., Boas, D.A. and Dunn, A.K. (2005). Determination of optimal exposure time for imaging of blood flow changes with laser speckle contrast

imaging. Applied Optics, [online] 44(10), p.1823.
doi:https://doi.org/10.1364/ao.44.001823.

167 Rajan, V., Varghese, B., Leeuwen, T.G. van and Steenbergen, W. (2009). Review of methodological developments in laser Doppler flowmetry. Lasers in Medical Science, [online] 24(2), pp.269–283. doi:https://doi.org/10.1007/s10103-007-0524-0.

168 Draijer, M., Hondebrink, E., van Leeuwen, T. and Steenbergen, W. (2008). Review of laser speckle contrast techniques for visualizing tissue perfusion. Lasers in Medical Science, [online] 24(4), pp.639–651. doi:https://doi.org/10.1007/s10103-008-0626-3.

169 Briers, D. (2007). Laser speckle contrast imaging for measuring blood flow. Optica Applicata, [online] 37(1-2). Available at: https://opticaapplicata.pwr.edu.pl/files/pdf/2007/no12/optappl_3712p139.pdf [Accessed 27 Oct. 2024].

170 Romero, G., Alanis, E. and Rabal, H.J. (2000). Statistics of the dynamic speckle produced by a rotating diffuser and its application to the assessment of paint drying. Optical Engineering, [online] 39(6), p.1652. doi:https://doi.org/10.1117/1.602542.

171 Boas, D.A. and Dunn, A.K. (2010). Laser speckle contrast imaging in biomedical optics. Journal of Biomedical Optics, 15(1), p.011109. doi:https://doi.org/10.1117/1.3285504.

172 Choi, B., Kang, N.M. and Nelson, J.Stuart. (2004). Laser speckle imaging for monitoring blood flow dynamics in the in vivo rodent dorsal skin fold model. Microvascular Research, 68(2), pp.143–146. doi:https://doi.org/10.1016/j.mvr.2004.04.003.

173 Choi, B., Ramirez-San-Juan, J.C., Lotfi, J. and Stuart Nelson, J. (2006). Linear response range characterization and in vivo application of laser speckle imaging of blood flow dynamics. Journal of Biomedical Optics, [online] 11(4), p.041129. doi:https://doi.org/10.1117/1.2341196.

174 EOS Data Analytics. (n.d.). Satellite Band Combinations: Analytical Methods For Imagery. [online] Available at: https://eos.com/make-an-analysis/ [Accessed 27 Oct. 2024].

175 Sharma, S. and Hashmi, M.F., 2022. Partial Pressure Of Oxygen. [online] StatPearls Publishing. Available at: https://www.ncbi.nlm.nih.gov/books/NBK493219/ (Accessed 22 December 2022).

176 Encyclopedia Britannica. (2010). Phosphorescence. [online] Available at: https://www.britannica.com/science/phosphorescence (Accessed 27 October 2024).

177 Roussakis, E., Spencer, J.A., Lin, C.P. and Vinogradov, S.A. (2014). Two-Photon Antenna-Core Oxygen Probe with Enhanced Performance. Analytical Chemistry, [online] 86(12), pp.5937–5945. doi:https://doi.org/10.1021/ac501028m.

178 Bolay, H., Reuter, U., Dunn, A.K., Huang, Z., Boas, D.A. and Moskowitz, M.A. (2002). Intrinsic brain activity triggers trigeminal meningeal afferents in a migraine model. Nature Medicine, [online] 8(2), pp.136–142. doi:https://doi.org/10.1038/nm0202-136.

179 Turbert, D. (2024) What is optical coherence tomography?, American Academy of Ophthalmology. Available at: https://www.aao.org/eye-health/treatments/what-is-optical-coherence-tomography (Accessed: 26 October 2024).

180 Fujimoto, J.G. et al. (2000) 'Optical coherence tomography: An emerging technology for biomedical imaging and optical biopsy', Neoplasia, 2(1–2), pp. 9–25. doi:10.1038/sj.neo.7900071.

181 Huang, D., Swanson, E., Lin, C., Schuman, J., Stinson, W., Chang, W., Hee, M., Flotte, T., Gregory, K., Puliafito, C. and et, al. (1991). Optical coherence tomography. Science, [online] 254(5035), pp.1178–1181. doi:https://doi.org/10.1126/science.1957169.

182 Venkatesh, R., Joshi, A. and Acharya, I. (2023) Optical coherence tomography biomarkers in diabetic macular edema, eOphtha. Available at: https://www.eophtha.com/posts/optical-coherence-tomography-biomarkers-in-diabetic-macular-edema (Accessed: 26 October 2024).

183 About plasmas and fusion (no date) Princeton Plasma Physics Laboratory. Available at: https://www.pppl.gov/about/about-plasmas-and-fusion (Accessed: 26 October 2024).

184 Nyayapathi, N. et al. (2024) 'Dual-modal photoacoustic and ultrasound imaging: From preclinical to clinical applications', Frontiers in Photonics, 5.

doi:10.3389/fphot.2024.1359784.

185 Yao, J. and Wang, L.V. (2013) 'Photoacoustic microscopy', Laser & Photonics Reviews, 7(5), pp. 758–778. doi:10.1002/lpor.201200060.

186 Pan, T. et al. (2021) Biophotonic probes for bio-detection and imaging, Nature News. Available at: https://www.nature.com/articles/s41377-021-00561-2 (Accessed: 26 October 2024).

187 Gather, M.C. and Yun, S.H. (2011). Single-cell biological lasers. Nature Photonics, [online] 5(7), pp.406–410. doi:https://doi.org/10.1038/nphoton.2011.99.

188 Catela, M. (2020). Biolaser: Concept and Applications. ResearchGate. [online] doi:https://doi.org/10.13140/RG.2.2.27142.40006.

189 Cartwright, J. (2011). A Cell Becomes a Laser. Science. [online] doi:https://doi.org/10.1126/article.28721.

190 Gordon Research Conferences (2024) Wikipedia. Available at: https://en.wikipedia.org/wiki/Gordon_Research_Conferences (Accessed: 26 October 2024).

191 Fan, X. and Yun, S.-H. (2014) 'The potential of Optofluidic biolasers', Nature Methods, 11(2), pp. 141–147. doi:10.1038/nmeth.2805.

192 Catela, M. (2020) (PDF) biolaser: Concept and applications, Biolaser: Concept and Applications. Available at: https://www.researchgate.net/publication/348383525_Biolaser_Concept_and_Applications (Accessed: 26 October 2024).

193 Haensch, T. (2005) (PDF) edible lasers: What's the next course?, Research Gate. Available at: https://www.researchgate.net/publication/286839746_Edible_lasers_What's_the_next_course (Accessed: 26 October 2024).

194 Optical cavity (2024) Wikipedia. Available at: https://en.wikipedia.org/wiki/Optical_cavity (Accessed: 26 October 2024).

195 Zhang, Z., Kimkes, T.E. and Heinemann, M. (2019) 'Manipulating rod-shaped bacteria with optical tweezers', Scientific Reports, 9(1). doi:10.1038/s41598-019-

55657-y.

196 Optical tweezers (2024) Wikipedia. Available at: https://en.wikipedia.org/wiki/Optical_tweezers (Accessed: 26 October 2024).

197 Berger, M. (2023) Optical tweezers explained, Nanotechnology. Available at: https://www.nanowerk.com/optical-tweezers-explained.php (Accessed: 26 October 2024).

198 Block lab at Stanford University (no date) Block lab - Optical tweezers. Available at: https://blocklab.stanford.edu/optical_tweezers.html (Accessed: 26 October 2024).

199 Urone, P.P. and Hinrichs, R. (2022) 29.4 photon momentum - college physics 2E, OpenStax. Available at: https://openstax.org/books/college-physics-2e/pages/29-4-photon-momentum (Accessed: 26 October 2024).

200 Handling nanoscale particles: 'next-generation' optical tweezers trap tightly without overheating (2011) ScienceDaily. Available at: http://www.sciencedaily.com/releases/2011/09/110926104132.htm (Accessed: 26 October 2024).

201 contact, P. (2011) 'next-generation' optical tweezers trap tightly without overheating, Harvard John A. Paulson School of Engineering and Applied Sciences. Available at: https://seas.harvard.edu/news/2011/09/next-generation-optical-tweezers-trap-tightly-without-overheating (Accessed: 26 October 2024).

202 Guck, J. et al. (2001) 'The Optical Stretcher: A novel laser tool to micromanipulate cells', Biophysical Journal, 81(2), pp. 767–784. doi:10.1016/s0006-3495(01)75740-2.

203 Mierke, C.T. (2019) 'The role of the optical stretcher is crucial in the investigation of cell mechanics regulating cell adhesion and motility', Frontiers in Cell and Developmental Biology, 7. doi:10.3389/fcell.2019.00184.

204 NCI Dictionary of Cancer terms (no date) Comprehensive Cancer Information - NCI. Available at: https://www.cancer.gov/publications/dictionaries/cancer-terms/def/extracellular-matrix (Accessed: 26 October 2024).

205 Endoscope (2024) Wikipedia. Available at:

https://en.wikipedia.org/wiki/Endoscope (Accessed: 26 October 2024).

206 Rehman , T. (2024) Endoscope vs. Borescope - what's the difference?, Ask Difference. Available at: https://www.askdifference.com/endoscope-vs-borescope/ (Accessed: 26 October 2024).

207 Understanding the difference between borescope and fiberscope use in aircraft maintenance (2023a) USA Borescopes. Available at: https://usaborescopes.com/news/understanding-the-difference-between-borescope-and-fiberscope-use-in-aircraft-maintenance/ (Accessed: 26 October 2024).

208 Terry, M. (2019) Experimental endoscope captures 3D images smaller than a single cell, BioSpace. Available at: https://www.biospace.com/new-endoscope-can-capture-3d-images-smaller-than-a-cell (Accessed: 26 October 2024).

209 Endoscope: Medlineplus medical encyclopedia (2023) MedlinePlus. Available at: https://medlineplus.gov/ency/article/002360.htm (Accessed: 26 October 2024).

210 Amer (2018) Endoscopy and types of endoscopy, SlideShare. Available at: https://www.slideshare.net/slideshow/endoscopy-and-types-of-endoscopy/117565504#5 (Accessed: 26 October 2024).

211 Reuter, M. (2006) [Philipp Bozzini (1773-1809): The endoscopic idealist]., Research Gate. Available at: https://www.researchgate.net/publication/6855299_Philipp_Bozzini_1773-1809_The_endoscopic_idealist (Accessed: 26 October 2024).

212 Rehman , T. (2024) Endoscope vs. Borescope - what's the difference?, Ask Difference. Available at: https://www.askdifference.com/endoscope-vs-borescope/ (Accessed: 26 October 2024).

213 Chavan, A. (2013) Introduction of laparoscopic surgery, SlideShare. Available at: https://www.slideshare.net/slideshow/introduction-of-laparoscopic-surgery/23902792 (Accessed: 26 October 2024).

214 Laparoscopy (2024) Johns Hopkins Medicine. Available at: https://www.hopkinsmedicine.org/health/treatment-tests-and-therapies/laparoscopy (Accessed: 26 October 2024).

215 Ridge, S.E., Shetty, K.R. and Lee, D.J. (2021) 'Current trends and applications

in endoscopy for Otology and neurotology', World Journal of Otorhinolaryngology - Head and Neck Surgery, 7(2), pp. 101–108. doi:10.1016/j.wjorl.2020.09.003.

216 Otology and neurotology: Conditions & treatments: UT southwestern medical center (no date) Conditions & Treatments | UT Southwestern Medical Center. Available at: https://utswmed.org/conditions-treatments/otology-and-neurotology/ (Accessed: 26 October 2024).

217 Brown, K. (2024) What is a mastoidectomy? types, indications, and more, WebMD. Edited by C. DerSarkissian. Available at: https://www.webmd.com/cold-and-flu/ear-infection/what-is-mastoidectomy (Accessed: 26 October 2024).

218 Ahmad, J.G., Shetty, K.R. and Alava, I. (2023) 'Advancements and innovations in otologic surgery: Endoscopic and Exoscopic Ear Surgery', Advancements and Innovations in OMFS, ENT, and Facial Plastic Surgery, pp. 63–77. doi:10.1007/978-3-031-32099-6_4.

219 Bailey, A. (2023) What is the mastoid?, Verywell Health. Available at: https://www.verywellhealth.com/mastoid-process-7496511 (Accessed: 26 October 2024).

220 Kutz, J.W. and Isaacson, B. (2019) Minimally invasive ear surgery offers significant advantages over open ear surgery: Aging: Prevention: UT southwestern medical center, Aging | Prevention | UT Southwestern Medical Center. Available at: https://utswmed.org/medblog/minimally-invasive-ear-surgery-offers-significant-advantages-over-open-ear-surgery/ (Accessed: 26 October 2024).

221 Madormo, C. (2023) Biopsy: What you need to know, Health. Available at: https://www.health.com/biopsy-7497693 (Accessed: 26 October 2024).

222 Zerbino, Dd. (1994). [Biopsy: its history, current and future outlook].. Likars'ka sprava / Ministerstvo okhorony zdorov'ia Ukraïny, 1-9.

223 Bigio, I.J. and Mourant, J.R. (n.d.) Encyclopedia chapter. Available at: https://web.mit.edu/hst.035/ReadingSupplement/05_03_Bigio/Bigio-Mourant_Encyclo_chap.pdf (Accessed: 27 October 2024).

224 Photodynamic therapy (2024) Mayo Clinic. Available at: https://www.mayoclinic.org/tests-procedures/photodynamic-therapy/about/pac-20385027 (Accessed: 26 October 2024).

225 PDT: What is PDT?: Photodynamic therapy (2021) PDT | What is PDT? | Photodynamic Therapy | American Cancer Society. Available at: https://www.cancer.org/cancer/managing-cancer/treatment-types/radiation/photodynamic-therapy.html (Accessed: 26 October 2024).

226 Overchuk, M. et al. (2023) 'Photodynamic and photothermal therapies: Synergy Opportunities for nanomedicine', ACS Nano, 17(9), pp. 7979–8003. doi:10.1021/acsnano.3c00891.

227 NCI Dictionary of Cancer terms (no date) Comprehensive Cancer Information - NCI. Available at: https://www.cancer.gov/publications/dictionaries/cancer-terms/def/antibody-drug-conjugate (Accessed: 26 October 2024).

228 Rodrigues, M.C. et al. (2022) 'An overview on immunogenic cell death in cancer biology and therapy', Pharmaceutics, 14(8), p. 1564. doi:10.3390/pharmaceutics14081564.

229 Kobayashi, H. and Choyke, P.L. (2019) 'Near-infrared photoimmunotherapy of cancer', Accounts of Chemical Research, 52(8), pp. 2332–2339. doi:10.1021/acs.accounts.9b00273.

230 RAO, M.C. (2013) 'Pulsed laser deposition — ablation mechanism and applications', International Journal of Modern Physics: Conference Series, 22, pp. 355–360. doi:10.1142/s2010194513010362.

231 What to expect at Laser Ablation surgery (2024) University of Utah Health | University of Utah Health. Available at: https://healthcare.utah.edu/neurosciences/neurosurgery/laser-ablation (Accessed: 26 October 2024).

232 Arkansas Children. (n.d.). ROSA ONE Brain Neurosurgery Technology | Arkansas Children's. [online] Available at: https://www.archildrens.org/programs-and-services/neuroscience-center/epilepsy-surgery/Treatments/rosa-one-brain [Accessed 27 Oct. 2024].

233 Seattle Children's (n.d.). Laser Ablation Surgery for Epilepsy and Brain Tumors. [online] Seattle Children's Hospital. Available at: https://www.seattlechildrens.org/clinics/neurosciences/services/laser-ablation-surgery-epilepsy-brain-tumors/ [Accessed 27 Oct. 2024].

234 University of Utah (2022). Laser Ablation Surgery. [online] University of Utah Healthcare. Available at: https://healthcare.utah.edu/neurosciences/neurosurgery/laser-ablation [Accessed 27 Oct. 2024].

235 Hashmi, J.T., Huang, Y.-Y., Osmani, B.Z., Sharma, S.K., Naeser, M.A. and Hamblin, M.R. (2010). Role of Low-Level Laser Therapy in Neurorehabilitation. PM&R, 2, pp.S292–S305. doi:https://doi.org/10.1016/j.pmrj.2010.10.013.

236 Ganipineni, V.D.P., Gutlapalli , S.D., Ajay Sai Krishna Kumar, I., Monica, P., Vagdevi, M. and Samuel Sowrab, T. (2023). Exploring the Potential of Energy-Based Therapeutics (Photobiomodulation/Low-Level Laser Light Therapy) in Cardiovascular Disorders: A Review and Perspective. Cureus, [online] 15(4). doi:https://doi.org/10.7759/cureus.37880.

237 Fast facts about Agriculture & Food (no date) American Farm Bureau Federation. Available at: https://www.fb.org/newsroom/fast-facts (Accessed: 26 October 2024).

238 United Nations (2013). World must sustainably produce 70 per cent more food by mid-century – UN report. [online] UN News. Available at: https://news.un.org/en/story/2013/12/456912 [Accessed 27 Oct. 2024].

238 Popp, J. (ed.) (2009) Journal of Biophotonics. Weinheim: Wiley-VCH.

239 d'Humières, B. (2023). Photonic Technologies for Agriculture. [online] SPECTARIS. Available at: https://www.spectaris.de/fileadmin/Infothek/Photonik/Zahlen-Fakten-und-Publikationen/2023_Photonic_Technologies_for_Agriculture.pdf [Accessed 27 Oct. 2024].

240 Polymerase chain reaction (2024) Wikipedia. Available at: https://en.wikipedia.org/wiki/Polymerase_chain_reaction (Accessed: 26 October 2024).

241 Poux, F. (2023) Representing 3D point cloud data, GIM International. Available at: https://www.gim-international.com/content/article/representing-3d-point-cloud-data (Accessed: 26 October 2024).

242 Friedli, M., Kirchgessner, N., Grieder, C., Liebisch, F., Mannale, M. and Walter,

A. (2016). Terrestrial 3D laser scanning to track the increase in canopy height of both monocot and dicot crop species under field conditions. Plant Methods, [online] 12(1). doi:https://doi.org/10.1186/s13007-016-0109-7.

243 The Definitive Guide to Slam & Mobile Mapping Technologies (no date) technologies. Available at: https://www.navvis.com/technology/slam (Accessed: 26 October 2024).

244 Specim. (n.d.). What is hyperspectral Imaging?: A Comprehensive Guide - Specim Spectral Imaging. [online] Available at: https://www.specim.com/technology/what-is-hyperspectral-imaging/ [Accessed 27 Oct. 2024].

245 Specim. (n.d.). Hyperspectral vs multispectral cameras: understanding advantages and limitations in spectral imaging. [online] Available at: https://www.specim.com/technology/hyperspectral-vs-multispectral-cameras/ [Accessed 27 Oct. 2024].

246 Hyperspectral Imaging in agriculture and vegetation (2024) Specim. Available at: https://www.specim.com/hyperspectral-imaging-applications/agriculture-and-vegetation/ (Accessed: 26 October 2024).

247 Sivarajah, Ilamaran. "Why Combine LiDAR and Hyperspectral Imaging?" AZoSensors, AZoSensors, 10 Nov. 2022, www.azosensors.com/article.aspx?ArticleID=2685. Accessed 30 Oct. 2024.

248 Phillips, K. (2023) Ken Phillips, Color Measurement Spectrophotometer Supplier & Manufacturer. Available at: https://www.hunterlab.com/blog/measuring-water-quality-with-spectrophotometry-the-best-approach-for-identifying-the-unknown/ (Accessed: 26 October 2024).

249 Historic trends in light intensity (2020) AI Impacts. Available at: https://aiimpacts.org/historic-trends-in-light-intensity/ (Accessed: 26 October 2024).

250 Spectrecology (2021). Difference & Similarities Spectrometer vs. Spectrophotometer. [online] Spectrecology. Available at: https://spectrecology.com/blog/spectrometer-vs-spectrophotometer/ [Accessed 27 Oct. 2024].

251 Diaz, S.O. and Viplav, A. (2024) Connection: OD, absorbance, and transmittance in spectrophotometry, byonoy.com. Available at: https://byonoy.com/journal/understanding-od-absorbance-transmittance-spectrophotometry/ (Accessed: 26 October 2024).

252 Guenther, Derek, and Harish Dahbi. "Optical Oxygen and PH Sensors for Monitoring Biofermentation Processes." BioProcess International, Informa Connect Division of Informa PLC, 1 Aug. 2011, www.bioprocessintl.com/process-monitoring-and-controls/optical-oxygen-and-ph-sensors-for-monitoring-biofermentation-processes. Accessed 30 Oct. 2024.

253 UV LED technology for emerging applications in agriculture (2018) LED professional - LED Lighting Technology, Application Magazine. Available at: https://www.led-professional.com/resources-1/articles/uv-led-technology-for-emerging-applications-in-agriculture (Accessed: 26 October 2024).

254 Mo, K. et al. (2024) 'Fresh meat classification using laser-induced breakdown spectroscopy assisted by LIGHTGBM and optuna', Foods, 13(13), p. 2028. doi:10.3390/foods13132028.

255 Moncayo, S. et al. (2016) 'Classification of red wine based on its Protected designation of origin (PDO) using laser-induced breakdown spectroscopy (LIBS)', Talanta, 158, pp. 185–191. doi:10.1016/j.talanta.2016.05.059.

256 Bilge, G. et al. (2016) 'Identification of meat species by using laser-induced breakdown spectroscopy', Meat Science, 119, pp. 118–122. doi:10.1016/j.meatsci.2016.04.035.

257 Kamruzzaman, M. et al. (2013) 'Fast detection and visualization of minced lamb meat adulteration using NIR hyperspectral imaging and Multivariate Image Analysis', Talanta, 103, pp. 130–136. doi:10.1016/j.talanta.2012.10.020.

258 Tian, X., Wang, J. and Cui, S. (2013) 'Analysis of pork adulteration in minced mutton using electronic nose of metal oxide sensors', Journal of Food Engineering, 119(4), pp. 744–749. doi:10.1016/j.jfoodeng.2013.07.004.

259 Ballin, N.Z., Vogensen, F.K. and Karlsson, A.H. (2009) 'Species determination – can we detect and quantify meat adulteration?', *Meat Science*, 83(2), pp. 165–174. doi:10.1016/j.meatsci.2009.06.003.

260 Wikipedia Contributors (2019). Biomimetics. [online] Wikipedia. Available at:

https://en.wikipedia.org/wiki/Biomimetics [Accessed 28 Oct. 2024].

261 Sofia, M.K. (2017). Moth Eyes Inspire Glare-Resistant Coating For Cellphone Screens. NPR. [online] 22 Jun. Available at: https://www.npr.org/sections/thetwo-way/2017/06/22/533803686/moth-eyes-inspire-glare-resistant-coating-for-cellphone-screens [Accessed 27 Oct. 2024].

262 Williams, M. (n.d.). Moth Eyes Inspire New Non Reflective Solar Panel Technology | HeroX. [online] HeroX. Available at: https://www.herox.com/blog/178-moth-eyes-inspire-new-means-of-boosting-solar-effi [Accessed 28 Oct. 2024].

263 Ayre, J. (2013). Brighter LEDs Inspired By Fireflies, Efficiency Increased By 55% | CleanTechnica. [online] CleanTechnica. Available at: https://cleantechnica.com/2013/01/09/brighter-leds-inspired-by-fireflies-efficiency-increased-by-55-percent/ [Accessed 28 Oct. 2024].

264 Wikipedia. (2023). Lunar Flashlight. [online] Available at: https://en.wikipedia.org/wiki/Lunar_Flashlight [Accessed 28 Oct. 2024].

265 Photo Solutions (2023). Photonics In Space: Small Components On Big Missions. [online] Photo Solutions. Available at: https://photo-solutions.com/photonics-in-space-small-components-on-big-missions/ [Accessed 27 Oct. 2024].

Images:

1 Venter, M. (2016) *Investigation and Characterization of the HERA Dish and Feed Using Electromagnetic Simulations*, *Research GAte*. Available at: https://search.datacite.org/works/10.13140/RG.2.1.1809.4807 (Accessed: 30 October 2024).

2 Total Internal Reflection. (2022). GeeksforGeeks. Available at: https://www.geeksforgeeks.org/total-internal-reflection/ [Accessed 21 Oct. 2024].

3 Our World in Data. (2023). Moore's Law: Transistors per microprocessor. [online] Available at: https://ourworldindata.org/grapher/transistors-per-microprocessor [Accessed 21 Oct. 2024].

4 *Dennis Gabor* (2024) *Wikipedia*. Available at:

https://en.wikipedia.org/wiki/Dennis_Gabor (Accessed: 30 October 2024).

5 Johnston, S.F. (2006) *Holography pioneers Emmett Leith (right) and Juris Upatnieks at the... | download scientific diagram, Research Gate*. Available at: https://www.researchgate.net/figure/Holography-pioneers-Emmett-Leith-right-and-Juris-Upatnieks-at-the-Radar-Optics-Lab_fig1_235222653 (Accessed: 30 October 2024).

6 Shoydin, S.A. and Pazoev, A.L. (2021) 'Transmission of 3D holographic information via conventional communication channels and the possibility of multiplexing in the implementation of 3D hyperspectral images', *Photonics*, 8(10), p. 448. doi:10.3390/photonics8100448.

7 *Geometrical optics* (2024) *Wikipedia*. Available at: https://en.wikipedia.org/wiki/Geometrical_optics (Accessed: 30 October 2024).

8 Magnification (2024) Wikipedia. Available at: https://en.wikipedia.org/wiki/Magnification (Accessed: 30 October 2024).

9 . Homogeneous circular polarizer. (n.d.). Wikipedia. Available at: https://en.wikipedia.org/wiki/Polarizer#cite_note-handedness-11 [Accessed 21 Oct. 2024].

10 Geethalakshmi, R. *et al.* (2022) 'Utilization of flow cytometry in nanomaterial/bionanomaterial detection', *Handbook of Microbial Nanotechnology*, pp. 133–144. doi:10.1016/b978-0-12-823426-6.00016-4.

www.ingramcontent.com/pod-product-compliance
Lightning Source LLC
Chambersburg PA
CBHW031630210526
45464CB00004B/1823